Professor C. B. Alcock

**Books are to be returned on or before
the last date below.**

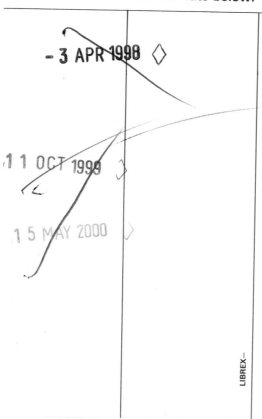

- 3 APR 1998 ◇

1 1 OCT 1999 ◇

1 5 MAY 2000 ◇

LIBREX—

THE CHARLES BENJAMIN ALCOCK SYMPOSIUM

# HIGH TEMPERATURE MATERIALS CHEMISTRY

Proceedings of the Symposium
Held at Imperial College, London,
28–29 October 1993
to Celebrate the 70th Birthday of
Professor C. B. Alcock

Edited by
B. C. H. STEELE
*Professor of Materials Science*
*Centre for Technical Ceramics*
*Department of Materials*
*Imperial College, London*

THE INSTITUTE OF MATERIALS

Book 586
First published in 1995 by
The Institute of Materials
1 Carlton House Terrace
London SW1Y 5DB

British Library Cataloguing-in-Publication Data

Charles Benjamin Alcock Symposium: High
Temperature Materials Chemistry -
Proceedings of the Symposium Held at
Imperial College, London, 28-29 October
1993 to Celebrate the 70th Birthday of
Professor C. B. Alcock
I. Steele, B. C. H.
620.11217

ISBN 0–901716–55–3

Typeset, printed and bound by
Bourne Press Ltd, Bournemouth, UK

# CONTENTS

# PREFACE
## Professor Charles Benjamin Alcock

To celebrate Professor 'Ben' Alcock's seventieth birthday on the 24th October 1993, two of his former postgraduate students Professors Paul Grieveson and Brian Steele, decided to organise a two-day meeting with lectures given at both Imperial College and the Geology Museum, South Kensington, London. The theme of the meeting was 'High Temperature Materials Chemistry', a topic which encompassed Professor Alcock's research interests from his early studies at BISRA, London, on FeS in 1947, to his current investigations on ceramic electrolyte sensors, and non-stoichiometric oxide perovskites, at the University of Notre Dame, Indiana. His distinguished research career has been recognised by many awards including the Kroll Medal, The Institute of Materials, UK and Paul Lebeau Medal (France). Twenty of these years were spent with Professor F.D. Richardson at the Nuffield Research Group for Extraction Metallurgy, Imperial College, followed by almost another twenty years at the University of Toronto. Whilst in Canada, 'Ben' became Chairman of the Department of Metallurgy and Materials Engineering, and was also elected to a Fellowship of the Royal Society of Canada. The organisers were delighted that so many of his former students and colleagues in North America were able to join others from Europe for the celebratory dinner at Imperial College, and to provide papers for the present volume.

Professor Alcock's contributions to both the thermodynamics and kinetics of high temperature materials' chemistry have covered a period which has witnessed a decline in the interest and financial support for this topic followed by a recognition over the past few years that acquisition of the relevant data and a better understanding of the associated technology is essential if materials processing and 'industrial ecology' are to meet the performance and environmental demands of modern society. Fortunately Professor Alcock's expertise and energy, whilst Chairman of the IUPAC Committee on High Temperature Chemistry and Materials, has helped to establish relevant computerised thermodynamic databanks, and it was also appropriate that just before his 70th birthday the 6th edition of *Metallurgical Thermochemistry* was published, and which commemorated his long association with the late Professor O. Kubaschewski.

Whilst still continuing investigations into Ostwald ripening and fuel cell materials at Toronto, Professor Alcock intends to find more time to indulge in his other interests, especially playing the piano, and enjoying the local wines of the Languedoc region during the extended visits which Valerie and he now make to their house in the south of France. We extend our very best wishes to them both and look forward to a continuing output of stimulating research papers and books.

B.C.H. Steele
Dept. of Materials
Imperial College

# Charles Benjamin Alcock

C.B.A. began his scientific career as a student of Chemistry in the Royal College of Science. After some post-graduate experience with neutron activation using a D–D generator which he brought into operation at University College, he joined BISRA in 1947.

Here, he developed a gas-recirculating apparatus in which he established the thermodynamics of FeS using $H_2/H_2S$ equilibration with radioactive sulphur. This equipment he then used to make the first quantitative study of thermal segregation in $H_2/H_2S$ mixtures.

After moving to the Royal School of Mines to join the Nuffield Research Group, he made his PhD study under the supervision of F. D. Richardson on the effects of alloying elements on the activity coefficient of sulphur in dilute solution in copper. Together they then developed atomic models of these alloys, based on the Regular solution and Quasi-chemical models to account for the thermodynamics. This was one of the earliest attempts to relate atomic models of ternary alloys to thermodynamics.

As a young research supervisor at Imperial College, he worked on the correct procedure for the measurement of vapour pressures by the gas-transportation and Knudsen techniques, and the use of solid electrolytes to measure oxygen potentials in the solid state and in dilute solutions of oxygen in liquid metals. As part of this latter study the technique known today as "coulometric titration" was developed for the first time. Other studies in this period (1955-1969) included diffusion measurements of uranium diffusion in $UO_2$ as a function of nonstoichiometry, the Ostwald ripening of thoria dispersed-phase nickel alloys, the kinetics of sulphation of some metal oxides and the kinetics of evaporation of ceramic oxides at temperatures around 2000K.

All of these kinetic studies required the development of new experimental methods such as torsion-effusion at 2000K, alpha-ray spectrometry for uranium and thorium diffusion, and high temperature cinematography for the observation of sulphation.

He left Imperial College to join the University of Toronto as Department chairman, in 1969, and continued his work largely using the techniques developed at Imperial College, to obtain data in some fields of then-current interest. These included further studies in ternary alloy systems, using now mass-spectrometry, the development of steelmaking oxygen probes with solid electrolytes, coulometric titration in the solid state, the thermodynamics of spinels, and model systems for the basic study of Ostwald ripening.

One entirely new area of research at Toronto, was the development of a novel and efficient plasma furnace. This began as a laboratory study, but was extended to the 1 Megawatt level in a nearby-pilot plant.

In 1987 he moved once again to the University of Notre Dame to direct a research group on electrochemical sensors. This group developed a new range of sensors, based on lanthanum and strontium fluorides. Sensors for oxygen from room temperature (in water) to 2000K in liquid metals, and sensors for a number of other elements were developed in the period 1987-1994. His group also collaborated with a group in catalysis to develop perovskite oxygen-deficient materials for elctrodes in fuel cells.

He has now returned to Toronto, in principle as a retired person, but at the present time is engaged in thermodynamic data assessment, Ostwald ripening studies in microgravity, and fuel cell materials development.

# Solutions in Progress

C. B. ALCOCK

*Freimann Chair Professor, University of Notre Dame, Notre Dame, Indiana 46556, USA*

## Retrospect

One of my earliest memories as a scientist in the field of high temperature chemistry was attending a very impressive conference held by the Faraday society in 1948. This was entitled 'The Physical Chemistry of Process Metallurgy', and amongst the speakers were Chipman, Ellingham, Pourbaix and a young F. D. Richardson, who was yet to begin his major work in forming the Nuffield Research Group in Extraction Metallurgy at the Royal School of Mines. At that time this field of research was the principal spawning ground of new ideas and experimental techniques in high temperature chemistry, and, of course, was mainly concerned with solution thermodynamics of liquid metallic alloys and slags. Solid state research was mainly confined to oxide ceramics, and was largely devoted to the study of phase diagrams.

It was not until 1961 that the burgeoning development of nuclear materials processing emerged into the general field of high temperature chemistry as a result of a conference held in Vienna under the auspices of the International Atomic Energy Agency. This conference, which was organised largely at the instigation of O. Kubaschewski, turned attention to a completely new group of metals, alloys and compounds which had only been studied in national laboratories allowing restricted access to their research and development. The nuclear field also turned attention to the physicochemical properties of ceramic materials in a much broader field than had been studied to that date and quantitative studies up to 2000°C soon became part of the general spectrum of experimental high temperature chemistry.

However, the world was not yet ready for the emergence of a subject called 'Materials Chemistry' with its implication of a predominating interest in solid state studies, and those of us taking part in this new development were labelled generally as 'Metallurgical Chemists'. It was under this title that I was appointed a Professor at Imperial College in 1964, and at my inaugural address I tried to bring together a perspective of what were the individual components of my subject and summarised these in the manner shown in Fig. 1. For those of you unfamiliar with the London underground system, the map I showed here reproduced the map of the Inner Circle Railway Line, representing the stations by a sequential naming in parallel with the connecting ideas fundamental to high temperature chemistry. On looking back at this map, I was struck by the fact that in those days we were concerned with a classical chemical approach to compounds and solutions in which the

1

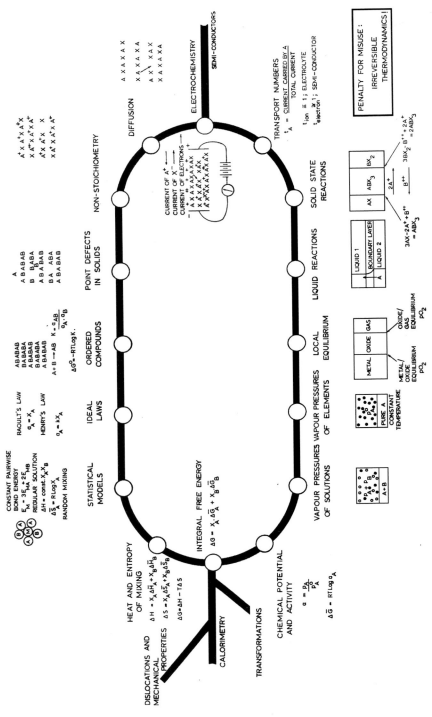

**Fig. 1** *Map of the Inner Circle in metallurgical chemistry.*

properties were measured in order to understand how processes, already largely in existence, could be understood quantitatively.

After this broad description of the field I discussed several experimental and theoretical aspects of a number of systems including semi-conductors and glass-ceramics which seemed to pose some 'Problems with Solutions' which was the title of my talk. I would like to review now some aspects of the developments during the subsequent thirty years which have led to the acceptance of the field of materials chemistry as a valid discipline. These are centred around solution chemistry, but nowadays that term includes non-stoichiometric solids as well as solid or liquid solutions.

## Alloy Thermodynamics

The arrival of the computer has greatly influenced our ability to assemble and test theoretical models against experimental data, and nowhere has this had a more profound effect than in the calculation of the properties of complex alloy systems from a knowledge of binary alloy data.

Darken's solution[1] of the ternary Gibbs–Duhem equation contained a suggestion that in a system $A$–$B$–$C$ where each component binary system $A$–$B$, $B$–$C$ and $A$–$B$ was strictly regular in behaviour, the Gibbs energy of formation of the ternary alloys over the whole composition range could be calculated quite simply from the data for each of the binaries.

Alcock and Richardson[2] tested this approximation to obtain the behaviour of a dilute constituent $C$, in a concentrated binary alloy, $A + B$, thus

$$\Delta \bar{G}_{C(A+B)} = X_A \Delta \bar{G}_{C(A)} + X_B \Delta \bar{G}_{C(B)} - \Delta G_{A+B}^{XS}$$

This equation holds quite satisfactorily for a number of systems involving metallic elements only, but fails when one component is highly electronegative such as oxygen or sulphur. We further developed an equation based on the quasi-chemical theory which took account of a large difference in the $C$–$A$ interaction from the $C$–$B$ interaction[3], but at the high coordination numbers associated with metallic alloys the quasi-chemical calculation differed very little from the simple regular solution equation. Despite this failure, a number of procedures have been evolved, mainly under the inspiration of the Calphad group,[4] for the calculation of ternary from binary data which have met with reasonable success for ternary and even more complex alloy systems. Obviously the great merit of these computational procedures is that they can, at least, indicate the composition ranges in which detailed laboratory measurements should be subsequently made, which would lead to economies in development costs.

I believe these procedures will be most useful when the interaction between the component elements of a metallic system are relatively weak, i.e. approximately Raoultian. However following Miedema's demonstration[5] of electron transfer effects, similar to Pauling's electronegativity effects[6] in alloy stabilities, it is to be expected that some metallic alloys will show strong negative departures from ideality due to these polarisation effects. In this case it is quite probable that departures from the simple behaviour similar to those which Richardson and I found for dilute solutions of oxygen and sulphur in binary metallic solvents, will appear in ternary and higher alloys systems.

In some recent mass-spectrographic studies which Butler and I carried out on dilute solutions of indium in Au–Cu, Pd–Cu and Au–Pd solvents,[7] we found that

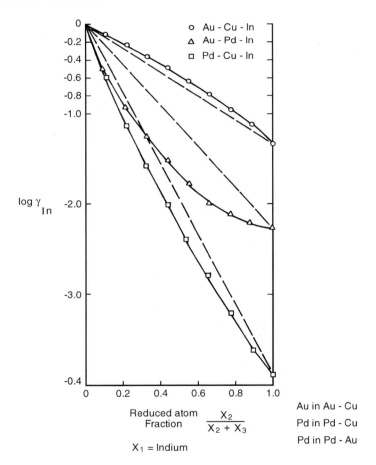

**Fig. 2** *The variation of the activity coefficient of indium in dilute solution in liquid Au–Cu, Au–Pd and Pd–Cu alloys at 1300 °C.*

the Au–Cu–In and Pd–Cu–In systems could be adequately accounted for by the regular solution approximation, but the Au–Pd–In system, in which both metals show a strong affinity for indium, showed departures which one could ascribe to electron transfer effects (Fig. 2). It is clear from the binary data that if one component, in this case copper, has a weak interaction with the dilute solute, the regular approximation will suffice.

These strong interactions are accompanied by significant departures from randomness, and therefore even when accurate heats of mixing and heat capacity data are available from modern calorimetry, such as that provided by the Tian–Calvet calorimeter, the question of the configurational entropy still remains unresolved. In order to obtain the total entropy of mixing data it is important to have complementary Gibbs energy data, either through direct activity measurements, or by the use of phase equilibrium information. Now that we have vapour pressure data for most of the elements,[8] it is possible to estimate the chance of success for activity measurements via vapour pressure measurements. It is clear that many alloy systems could not be measured accurately using this method in a reasonably accessible range of temperatures. It would seem that developments in electrochemical measurement capabilities are still needed to fill this gap. The range and flexibility of solid electrolytes for use in this field is demonstrated by the contributions of Professors Weppner and Fray to this Symposium.

## The Fabulous Perovskites

In the recent decade a great deal of interest has centred on the perovskite and closely related structures in the development of a wide range of electroceramics. The members of what I call the 'Perovskite Family Tree' (Fig. 3) provide examples of superconductors, semiconductors, solid electrolytes and dielectric materials. The chemistry of the systems derived from simple precursors with this structure has a wide-range flexibility due to the wide range of ions of constant or variable valency which can be substituted on the large cation A sites or the small cation B sites while retaining the same structure. This wide range of potential materials for applications as electrical materials exemplifies a growing point in the development of materials chemistry, in which precursors are varied by ionic substitution in order to produce materials having a specific range of properties suitable to a specific range of applications.[9] The inorganic chemist is no longer satisfied, as in the classical procedure, to prepare and characterise materials and test the products for a given application. The new thrust is similar to the *modus operandi* of the synthetic organic chemist in which atoms are brought together in various structural arrangements in order to produce a substance with desired properties. We have therefore moved from materials preparation to materials design as a major effort.

This synthetic approach is no better illustrated than in the search for catalysts to replace the platinum metals in a number of industrial and domestic applications. The group of perovskites based on the lanthanum compounds $LaMO_3$ (M = Cr, Mn, Fe, Co, Y) show a wide range of response to variation in the oxygen potential of the surrounding atmosphere in the temperature range 500–1000°C. The first four compounds can be made to have a high oxygen ion diffusion coefficients, when some lanthanum ions are replaced by strontium, at low oxygen potentials. At high oxygen pressures, around one atmosphere, the oxygen vacancies which are produced by Sr for La substitution are filled, and the valency of the transition metal ion is raised, leading to a high level of semiconduction. $La_{1-x}Sr_xYO_{3-x/2}$ remains a solid electrolyte over a wide range of oxygen because $Y^{3+}$ ion cannot be readily

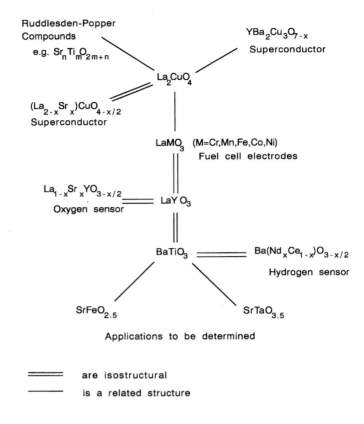

**Fig. 3**  *The perovskite family tree.*

oxidised to $Y^{4+}$. In a study of the catalytic effect for CO and $CH_4$ oxidation[10,11] it was shown that the transition metal compounds with Cr, Mn, Fe and Co function as good catalysts for CO oxidation, but the solid electrolyte based on $Y^{3+}$ is the most effective catalyst for $CH_4$ oxidation (Fig. 4). The upper figure, which shows the variation of the stoichiometry of the transition metal perovskites as a function of oxygen pressure, demonstrates that the chromium compound will have a high positive hole conduction over a wide oxygen pressure range, whereas the cobalt compound is readily reduced. The fact that the line for this compound and for the manganese compound cross the line

$$\delta = x/2$$

shows that reduction occurs to the divalent state at the low oxygen pressure. The iron compound has only $Fe^{3+}$ ions down $10^{-14}$ atmospheres, similar to the $Mn^{3+}$ compound, before $Fe^{2+}$ ions are produced.

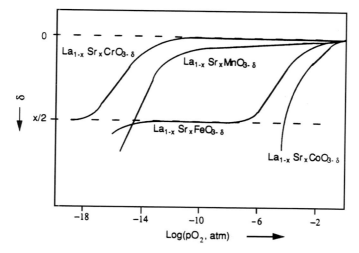

**Fig. 4 a.** *The non-stoichiometry of some Perovskites at 1000 °C.*

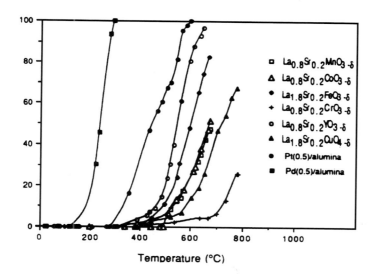

**Fig. 4 b.** *Methane oxidation on A-site-substituted perovskite catalysts.*

It appears that the catalytic activity for CO oxidation requires a high positive hole conductivity in the catalyst with the oxygen diffusion being of lesser importance. For $CH_4$ oxidation, the reverse seems to be the case, but the superior behaviour of the electrolyte $La_{1-x}Sr_xYO_{3-x/2}$ may also be related to the larger size of the $Y^{3+}$ ion playing a role in the adsorption of methane.

Aside from these substituted $A^{3+}B^{3+}O_3$ perovskites the $A^{2+}B^{4+}O_3$ materials have been, until recently, mainly of interest as dielectric materials for capacitors. The

7

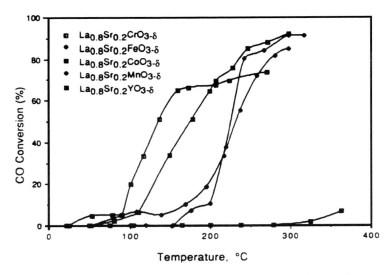

**Fig. 4 c.** *Carbon monoxide oxidation on A-site-substituted perovskite catalysts.*

recent development of a group of proton-conducting electrolytes such as the substituted barium cerate shown on the tree have opened a new phase in solid state electrochemistry. Both substituted cerates and zirconates have been employed, but it seems that the major centre of interest revolves around the Group III element which is substituted on the $B^{4+}$ sites. The mechanism of proton conduction arises because of the accompanying oxygen vacancies which are introduced. These will absorb oxygen from water vapour and, together with the accompanying hydrogen atoms, lead to hydroxyl group formation.[12] These groups give rise to the proton-conduction. There will clearly be a cross-over where the electrolytic conduction mechanism changes from proton to oxygen ion conduction, and this will depend on the temperature. At elevated temperatures the water solubility, which gives rise to proton conduction, diminishes considerably, whereas the oxygen ion mobility increases. The contribution of proton conduction to the electrolytic behaviour will be therefore determined by the extent to which the Group III B-site element affects the water solubility at a given temperature for a number of related systems. It might be expected that the relative stabilities of metal hydroxides would give some guidance in this respect, but the data for hydroxides are sparse and largely unreliable at the present time.

A number of compounds are closely related to the perovskite structure. The superconductors which have been the object of the most intense, and expensive, research effort, and the largely unexplored Ruddlesden–Popper compounds are related to perovskite through the interleaving of perovskite with rocksalt unit cells. At the other end of the tree the derivatives in which the 4+ cation is replaced by a 3+ cation (brownmillerite structure) show the presence of 'tunnels' where strings of oxygen ions are missing, and the substitution of a 5+ ion such as in $Sr_2Ta_2O_7$ leads to slabs of perovskite separated by defect slabs, somewhat similar to β alumina.[13] The two aliovalent substitutions should lead to novel physical and electrical properties which have so far not been developed.

## Surface Chemical Effects

Studies of catalysis by the platinum-group elements have shown that, as might be expected, the best results are obtained when the surface/volume ratio of the catalyst is maximised. This can best be achieved by the preparation of, for example, alumina-supported catalysts in which the metal catalyst is in the form of fine particles dispersed on the surface of the support. These materials show the deleterious effects of sintering by surface migration in a manner exactly similar to the process of Ostwald ripening by volume diffusion in disperse-phase strengthened alloys.

In an attempt to circumvent this phenomenon, a good deal of attention has been recently paid to the use of oxide catalysts as discussed above which do not undergo significant sintering at moderate temperatures, but which have the high electronic conductivity and oxygen-insensitive properties shown by the platinum group metals. In order to optimise the surface/volume ratio in these materials, sol-gel preparation methods have been used to prepare fine particle size catalysts. Materials of the perovskite compositions $(La_{1-x}Sr_x)MO_{3-\delta}$ (where M = Cr, Mn, Fe, Co) have been found to perform the catalyst role satisfactorily with surface areas of about 1 m²/g when compared with $Pt/Al_2O_3$ catalysts having specific surfaces several hundred times larger. The oxide solid solutions have a very large compositional variability which cannot be matched in alloy systems, and can be prepared to have $p$-type or $n$-type conductivity under a range of oxygen potentials and a wide range of temperatures.

The ceramics technologist now has a wide range of preparation techniques,[14] including the recently developed nano-phase preparation by condensation from a reactive atmosphere. This procedure, which has been used to prepare materials with an average particle size of 10–20 Å, has produced some surprises by way of physical properties, but needs to be further developed to make possible the production of the sophisticated compositions referred to above. It is clear, however, that once much higher specific surfaces of oxide solid solutions can be achieved, these materials will present a significant challenge to metal and alloy catalysts in terms of surface stability and chemical flexibility.

The failure of supported noble metal catalysts because of crystal growth via surface diffusion is usually ascribed to the phenomenon now known as 'Ostwald ripening'.

The classical analysis of Ostwald ripening in condensed phase systems by Wagner,[15] and Lifshitz and Slyozov[16] proposed that the mechanism of ripening could be rate-controlled by the kinetics of interface transfer between the dispersed phase and the matrix according to the equation

$$r^2(t) - r^2(o) = KC_oV_{M\gamma}/81RT$$

where $r(o)$ and $r(t)$ are the average dispersoid radii at time zero and $t$ respectively, $C_o$ is the solubility of a planar sample of dispersoid, $V_M$ and $\gamma$ are the molar volume and surface energy of the dispersoid respectively (Fig. 5).

9

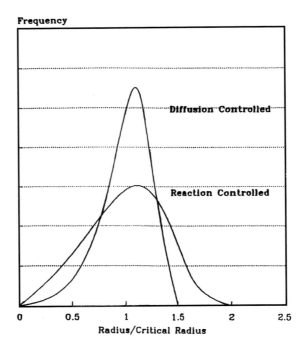

Ostwald ripening particle size distribution equations.

Interface transfer control

$$r^2(t) - r^2(0) = KC_oV_m\gamma/81RT$$

Matrix diffusion control

$$r^3(t) - r^3(0) = kC_oV_m\gamma/9RT$$

**Fig. 5** *Theoretical curves for particle size distribution as a result of Ostwald Ripening.*

The alternative mechanism, diffusion control by dispersoid atom transfer in the matrix, is rate-controlled according to

$$r^3(t) - r^3(o) = kC_oDV_{M\gamma}/9RT$$

where $k$ is a numerical constant and $D$ is the relevant diffusion coefficient.

These two equations lead to characteristic particle size distributions as a function of time and should provide clear criteria. However, there are a number of problems associated with these derivations, two of which arise from the initial particle size distribution and the value of the interfacial energy. In practical circumstances when dispersed phase alloys are made the particle size distribution is frequently characterised by bimodality, and by a long tail towards the larger end of the size distribution. This tail did not disappear during most experimental studies, a result not described by the theoretical equations. The second factor, the interfacial energy, is rarely known for the solid–solid systems which have been studied. In one example where the interfacial energy is reasonably well known, the growth of voids in metallic matrices, the tail of the distribution still cannot be accounted for by the theoretical equations. One further experimental difficulty with solid–solid systems is that the counting statistics are usually poor because of the difficulty in measuring particle size distribution in solid samples. The results of these studies frequently do not show one particle size radius dependence clearly.

Kuczynski[17] has suggested that the problem with these two equations is that the real controlling factor in Ostwald ripening is the minimum rate of entropy production and this leads to a significantly different particle size distribution than either of the previous suggestions, and does include a long 'tail' at the large particle end of the distribution.

## The Power Crunch

During the 30 years since my inaugural lecture, nuclear power has, apparently, passes through its zenith, and is now labouring under the spectre of Chernobyl. A small but vociferous anti-nuclear movement is campaigning against the nuclear option for energy to such an extent that the leaders of the main industrial countries, except France, seem to have turned their backs on further developments. These include even those which could help in the containment of fission products such as the proposed Integral Fast Reactor.[18] This important concept, which was developed at the Argonne National Laboratory, has the fuel advantages of breeding reactor systems together with a fuel recycling circuit which is built adjacent to the reactor and which can return actinides to the fuel cycle. There is also a significant potential for fission product concentration to improve safety in disposal. A further nuclear project which has been proposed by Los Alamos National Laboratory would study the bombardment of fission products with high energy protons or similar charged particles with the objective of restoring the proton/neutron ratio to a value where stable nuclei would replace the radioactive fission product nuclei. This offers hope of removing permanently the need to store nuclear by-products under maximum security to eliminate the health risk. In this context, it is astonishing to the concerned scientist to read in President Clinton's first 'State of the Union' message that, 'We are eliminating programs that are no longer needed, such as nuclear power research and development.' If this ill-informed point of view is allowed to determine the future course of events, other energy transformation systems must be developed with renewed vigour. Of the two main options which are currently receiving significant funding and research effort, the battery-driven automobile seems to present a very limited contribution to the solution of the problem in the near future. The fuel cell is, however, undergoing a renaissance, and a number of new options, such as those described by Professor Steele, are in a promising stage of development.

At Notre Dame we have applied the perovskite materials which we used in our catalyst studies as electrodes on the highly conductive $Bi_2O_3$–$SrO$ and $CeO_2$–$CaO$ solid electrolytes in order to study oxygen transfer through the pseudo-fuel cell arrangement,

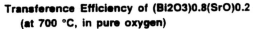

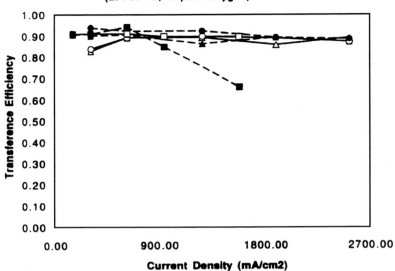

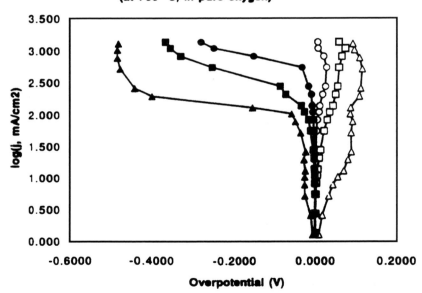

(Fig. 6)

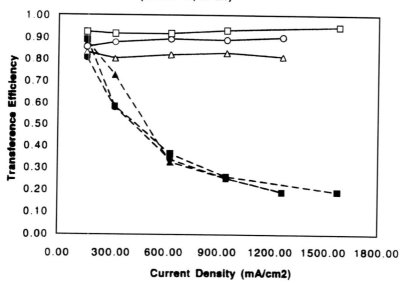

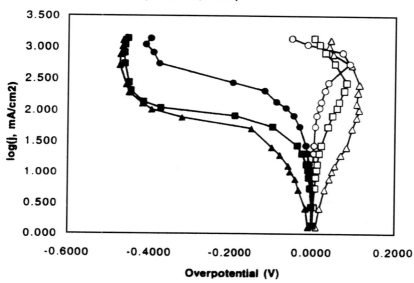

**Fig. 6** *Transference efficiency of oxygen at 700°C and overpotential at anode and cathode of $(Bi_2O_3)_{0.8}(SrO)_{0.2}$ solid electrolyte.* —△— *anodic, Pd/Au;* --▲-- *cathodic, Pd/Au;* —○— *anodic, La0.7Sr0.3MnO3-δ;* --●-- *cathodic, La0.7Sr0.3MnO3-δ;* —□— *anodic, La0.7Sr0.3FeO3-δ;* --■-- *cathodic, La0.7Sr0.3FeO3-δ.*

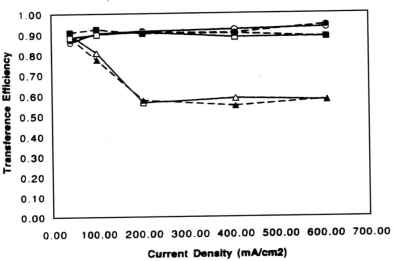

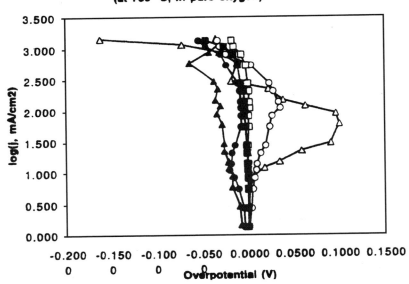

(Fig. 7)

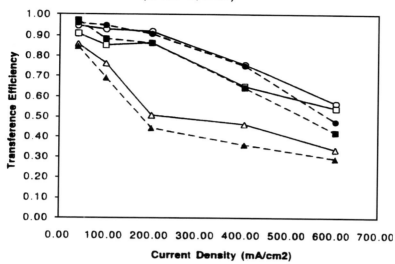

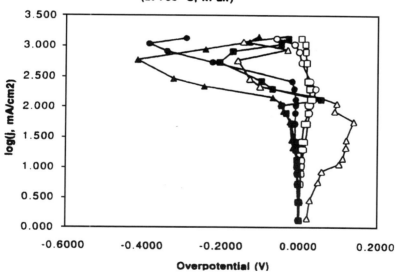

**Fig. 7** *Transference efficiency of oxygen at 700°C and overpotential at anode and cathode of $(CeO_2)_{0.1}$ solid electrolyte. —△— Pd/Au, anodic; —▲— Pd/Au, cathodic; —○— La0.7Sr0.3MnO3-∂, anodic; —●— La0.7Sr0.3MnO3-∂, cathodic; —□— La0.7Sr0.3FeO3-∂, anodic; —■— La0.7Sr0.3FeO3-∂, cathodic.*

O2  |  electrode  |  solid electrolyte  |  electrode  |  $O_2$

where the partial pressure of oxygen was the same at each electrode, and oxygen was transferred from the cathode to the anode by the passage of current.[19] The overvoltage at each electrode and the transfer efficiency, which is the volume of oxygen removed or delivered, divided by the volume calculated from the number of coulombs which were passed during an experiment, were measured. The results for the manganese and iron-based perovskite electrodes are shown in Figs 6 and 7.

It was found that the main source of overpotential and the lowest transfer efficiencies occurred at the cathode and when air was used as a source of oxygen. It is clear from these studies that the interfacial conditions at the cathode gas/electrode/electrolyte assembly are rate determining and that this is the central problem for the development of successful fuel cells at operating temperatures which are significantly less than 1000°C. There is clear evidence from these data that at high current densities, around 100 mA/cm² and higher, the $Bi_2O_3$-based electrolyte is reduced to bismuth metal and therefore that the *electrolyte* must supply the deficiency in the gas/electrode supply to maintain high current levels in our arrangement.

## Basic Science Funding

There have been a number of changes, most of them unfortunate, in the funding of basic research in materials chemistry during the last 30 years. This period has seen the rise and demise of many industrial laboratories, probably since the profitability of such operations to the production of industry is the final touchstone of success. There was a brief period when it was generally believed that profitability could be product-led rather than by a reduction in production costs which is the main objective today. The funding available to the remaining university centres of basic research has also been allowed to dwindle, largely in the naive belief that industry would supplement the shortfall in government support, and in that role, would direct 'basic research' towards the improvement of the national economy. Knowing as we do how basic research advances by 'fits and starts' and how this year's unfashionable area can provide next year's 'hot topic', it is clear to us in the field that like the Church and the State, basic research and industry are better separated. We can only hope that a future cadre of scientists with political acumen, but not 'political scientists' will give governments guidance as to what is to be done, and so replace the industrial leaders who have the ear of government at present.

## The Giants of Materials Chemistry

It is fitting to end this review of some aspects of progress in inorganic solution science which constitutes the major segment of materials chemistry by paying tribute to some of the leading scientists whose work has brought us to the present state. I have presented a list of names, in alphabetical order, and not necessarily of merit, which contains many men whom I was privileged to know personally (Fig. 8). All of them have made significant scientific advances, and the larger number of them

"If I have seen further, it is by standing on the shoulders of Giants."
Newton. 1675

| | | |
|---|---|---|
| J. S. Anderson | J. Benard | J. Chipman |
| L. S. Darken | H. J. T. Ellingham | H. Flood |
| G. Gerassimov | W. Hume-Rothery | W. Jost |
| O. Kubaschewski | J. Lumsden | A. Magneli |
| W. Olsen | H. Nowotny | M. J. N. Pourbaix |
| F. D. Richardson | H. Schenck | T. Takahashi |
| A. E. Van Arkel | C. Wagner | |

**Fig. 8** *Some Giants of Materials Chemistry.*

have left behind flourishing schools which now lead the way to new knowledge. Their life's work has built a firm foundation from which our present knowledge has sprung and if there were a 'Hall of Fame' in materials chemistry these men would certainly form the core membership.

## Prospect

The Royal Society of Chemistry has recently introduced a *Journal of Materials Chemistry* which recognises the maturity of this subject and the particularly interdisciplinary nature of this field of research. The journal coverage will be quite familiar to workers in materials chemistry, and together with journals such as *Solid State Ionics* and the recently broadened scope of many of the traditional metallurgical journals, should provide for adequate growth potential of the literature of our subject.

The experimental techniques which can be used to study the physical chemistry of inorganic materials has also reached a level of maturity where many options are available. It would appear that there is an increasing dearth of experimentalists to use these tools, and much greater effort is now being directed towards computer modelling of processes, or the computerised analysis of extant data. While this is not wholly deplorable, we can only hope that the next wave of optimism in the fortunes of the advanced countries will lead to the arrival of a new generation of chemists who will work with materials directly, and not only through the literature.

One growing area is in process control on the industrial scale. Here there is a great opportunity for sensor development, and whereas it has been difficult to persuade industry to use sensors as part of their everyday operations in materials processing, I have the impression that this resistance is being overcome by the arrival of a new generation of industrial chemists to whom the use and advantages of computers is a matter of second nature.

It is becoming clear too that the subtleties of structure in both stable and metastable equilibrium are becoming slowly apparent, and that there is need for advance in the everyday understanding and use of many physical techniques apart from the basic diffraction techniques, and especially with regard to surfaces. Surface composition and morphology both on free surfaces and at interfaces has long been a field of intense study in metal systems but much more can, and should, be done in the more complex field of non-metallic materials.

Finally, the new opportunities which sample preparation and high temperature experimentation under microgravity present should be pursued. This is especially so in the case of liquid systems in which reactions are occurring, such as during the electrolysis of molten salts, and in the preparation of dispersed or generally composite systems where the absence of gravity permits the mixing of phases of significantly differing density. Dr Das Gupta will present a brief description of our joint study in microgravity of two-phase systems during Ostwald ripening, which indicates some of the advantages and difficulties of working in the space environment.

Finally, let me thank all of my friends who have taken the time to come to this Symposium; their presence has given me a great deal of pleasure, and their scientific contributions which are gathered in the volume should give a broad example of the fascination of the field in which we have all worked.

## References

1. L. S. DARKEN: 'Application of the Gibbs-Duhem Equation to Ternary and Multicomponent Systems,' *J. Am. Chem. Soc.,* 1950, **72**, 2909–2914.

2. C. B. ALCOCK and F. D. RICHARDSON: 'Dilute Solutions in Molten Metals and Alloys,' *Acta Met.,* 1958, **6**, 385–395.

3. C. B. ALCOCK and F. D. RICHARDSON: 'Dilute Solutions in Alloys,' *Acta Met.,* 1960, **8**, 882–887.

4. I. ANSARA: 'Prediction of Thermodynamic Properties of Mixing and Phase Diagrams in Multi-component Systems,' *Metallurgical Chemistry Symposium 1971,* O. Kubaschewski ed., H.M. Stationery Office, London, 1972.

5. A. R. MIEDEMA, P. F. de CHATEL and F. R. de BOER: 'Cohesion in Alloys-Fundamentals of a Semi-empirical Model,' *Physica,* 1980, **100B**, 1–28.

6. L. PAULING: *The Nature of the Chemical Bond, 2nd Edition,* Cornell University Press, New York, 1952.

7. C. B. ALCOCK and J. BUTLER: 'Strong Interactions in Ternary Alloys,' Submitted to *Journal of Alloys and Compounds.*

8. C. B. ALCOCK, V. ITKIN and M. HORRIGAN: 'The Vapour Pressures of the Metallic Elements,' *Can. Met. Quart.,* 1984, **23**, 309–313.

9. C. B. ALCOCK: 'Thermodynamic and Transport Properties of Electroceramic Oxide Systems,' *J. Alloys and Compounds,* 1993, **197**, 217–227.

10. R. DOSHI, C. B. ALCOCK, N. GUNASEKARAN and J. J. CARBERRY: 'Carbon Monoxide and Methane Oxidation Properties of Oxide Solid Solution Catalysts,' *J. of Catalysis,* 1993, **140**, 557–563.

11. R. DOSHI, C. B. ALCOCK and J. J. CARBERRY, 'Effect of Surface Area on CO Oxidation by the Perovskite Catalysts $La_{1-x}Sr_xMO_{3-\delta}$ (M = Co, Cr),' *Cat. Letters,* 1993, **18**, 337–343.

12. T. NORBY: 'Proton Conduction in Oxides,' *Solid State Ionics,* 1990, **40/41**, 857–862.
13. D. M. SMYTH: 'Defect and Order in Perovskite-related Oxides,' *Ann. Rev. Mat. Sci.,* 1985, **15**, 329–357.
14. D. SEGAL: *Chemical Synthesis of Advanced Ceramic Materials,* Cambridge University Press, Cambridge, 1989.
15. C. WAGNER: 'Theorie der Altung von Niederschlagen durch Umlosen,' *Zeit fur Elektrochem,* 1961, **65**, 581–591.
16. M. LIFSHITZ and V. V. SLYOZOV: 'The Kinetics of Precipitation from Supersaturated Solid Solutions,' *Phys. Chem. Solids,* 1961, **19**, 35–50.
17. G. C. KUCZYNSKI: 'Minimum Entropy Production and Ostwald Ripening,' *Sintering Processes,* G. C. Kuczynski ed., Plenum Publishing Corp., NY, 1980.
18. Y. I. CHANG: 'The Integral Fast Reactor,' *Nucl. Tech,* 1989, **88**, 129–138.
19. R. DOSHI, Y. SHEN and C. B. ALCOCK: 'Oxygen Pumping Characterics of Oxide Ion Electrolytes at Low Temperatures,' Submitted to *Solid State Ionics.*

# Phase Stability in the Systems $YBa_2Cu_3O_{7-x}$, $Nd_2CuO_{4-\delta}$ and $Nd_{1.85}Ce_{0.15}CuO_{4-\delta}$

D. R. GASKELL AND Y. S. KIM

*School of Materials Engineering, Purdue University, West Lafayette, Indiana, USA*

## Abstract

The dependencies of the degrees of oxygen non-stoichiometry in the systems $YBa_2Cu_3O_{7-x}$ (123), $Nd_2CuO_{4-\delta}$ and $Nd_{1.85}Ce_{0.15}CuO_{4-\delta}$ on temperature and oxygen pressure have been determined in the range of temperature 350–950°C and the range of oxygen pressure $10^{-6}$ - 1 atm. The nature of the decomposition of these compounds has been determined by thermogravimetric analysis and X-ray diffraction. With decreasing oxygen pressure in the temperature range 750–950°C, 123 follows the sequence $123 \rightarrow Y_2BaCuO_5$ (211) $+ BaCuO_2 + Cu_2O \rightarrow 211 + BaCuO_2 + BaCu_2O_2 \rightarrow 211 + YBa_3Cu_2O_x$ (132) $+ BaCu_2O_2 \rightarrow 211 + BaCu_2O_2 + BaO$. Decreasing the oxygen pressure causes $Nd_2CuO_{4-\delta}$ to decompose in the sequence $Nd_2CuO_{4-\delta} \rightarrow NdCuO_2 + Nd_2O_3 \rightarrow Nd_2O_3 + Cu_2O$, and the corresponding decomposition in $Nd_{1.85}Ce_{0.15}CuO_{4-\delta}$ is $Nd_{1.85}Ce_{0.15}CuO_{4-\delta} \rightarrow Nd_2O_3 + NdCeO_{3.5} + Cu_2O$. Quantitative phase stability diagrams have been constructed.

## Introduction

The oxidation state of copper in $YBa_2Cu_3O_{7-x}$, (123), maintained in an oxygen atmosphere, varies from $x = 1$ at the highest temperature of approximately 1010°C to $x = 0$ at approximately 400°C, and several thermogravimetric studies of the influence of temperature and oxygen pressure on the non-stoichiometry of 123 have been conducted.[1-6] The dependence of the decomposition temperature of 123 on oxygen pressure has also been studied by several groups using thermogravimetric analysis[7] and galvanic cells.[8-10] Bormann and Nölting.8 observed that 123 decomposes at almost the same conditions as are required for the reduction of CuO to $Cu_2O$. In contrast, Lindemer *et al.*[7] and Ahn *et al.*[9] observed that the decomposition of 123 occurs at oxygen pressures significantly lower than those required for the transition of CuO to $Cu_2O$, although the results of these two studies are in disagreement with one another. Disagreement also exists as to the nature of the decomposition products of 123. Bormann and Nölting[8] and Porat *et al.*[10] have reported that the products of decomposition are $211 + BaCuO_2 + Cu_2O$, and Lindemer *et al.*[7] and Ahn *et al.*[9] identified the products of decomposition as $211 + BaCu_2O_2 + YBa_3Cu_2O_x$, (132).

The phase relations occurring in the pseudo-ternary system $Nd_2O_3$–$CeO_2$–CuO at 1000°C in atmospheric air have been studied by Pieczulewski *et al.*[11] who observed that large concentrations of cerium cause the formation of additional phases CuO and/or $CeO_2$. They found that the maximum solubility of Ce occurs at $x = 0.2$ in

21

$Nd_{2-x}Ce_xCuO_{4-\delta}$, which is the only ternary phase occurring in the quaternary system. A partial phase diagram for the system $Nd_2O_3$–$CuO$, determined by Oka and Unoki,[12] shows that $Nd_2CuO_4$ melts incongruently at about 1240°C and shows a eutectic at 1040°C between $Nd_2CuO_4$ and $CuO$.

Controversy exists as to whether or not $Nd_{2-x}Ce_xCuO_4$ can contain more than four oxygens.[13–15] The main difficulties in determining the oxygen content of the system arise from:

1. difficulty in the preparation of a specimen of $Nd_{2-x}Ce_xCuO_4$ which does not contain separate phases such as $CuO$, $CeO_2$ and phases in the binary system Nd–Cu;

2. $Nd_{2-x}Ce_xCuO_4$ exhibits internal inhomogeneity;

3. the valence of Ce may vary from +3 to +4, and the change in valence, which may occur at low oxygen pressures, has not been clearly identified.

The oxygen non-stoichiometry in $Nd_2CuO_{4-\delta}$ and $Nd_{1.85}Ce_{0.15}CuO_{4-\delta}$ has been studied in the range of oxygen pressure $10^{-4}$ - 1 atm by Suzuki *et al.*[15] using thermogravimetry, and Petrov *et al.*[16] have used EMF cells to study the phase relationships in the systems Ln–M–O, where Ln = La, Pr, Nd and M = Co, Ni, Cu.

In the present study, oxygen non-stoichiometry and the phase stability diagrams for the systems 123, $Nd_2CuO_{4-\delta}$ and $Nd_{1.85}Ce_{0.15}CuO_{4-\delta}$ have been determined using thermogravimetry, differential thermal analysis and X-ray diffraction.

## Experimental Procedures

Specimens of 123 were prepared by sintering mixtures of $Y_2O_3$, $BaCO_3$ and $CuO$ and specimens of $Nd_2CuO_{4-\delta}$ and $Nd_{0.85}Ce_{0.15}CuO_{4-\delta}$ were prepared by sintering mixtures of the oxides $Nd_2O_3$, $CeO_2$ and $CuO$. The sintering procedure involved heating cold-pressed pellets of the mixtures of powers at temperatures in the range 910–1080°C in flowing oxygen at 1 atm pressure for 48 h before cooling to room temperature. The pellets were then reground and cold-pressed to form new pellets which were then heated as before. This cycle was repeated three or four times until powder X-ray diffraction analysis showed that single phases had been produced.

Measurements of the oxygen non-stoichiometry of 123 were made by thermogravimetric analysis in the temperature range 400–950°C in flowing oxygen–nitrogen mixtures with partial pressures of oxygen in the range $10^{-6}$ - 1 atm, and the oxygen non-stoichiometries of $Nd_2CuO_{4-\delta}$ and $Nd_{1.85}Ce_{0.15}CuO_{4-\delta}$ were made in the range 350–950°C with partial pressures of oxygen in the range $10^{-5}$ - 1 atm. A piece of pellet of approximate dimensions 1mm x 1mm x 1mm was placed on a thin alumina disk which was suspended from the arm of a Du Pont TGA balance and the partial pressure of oxygen in the gas flowing through the TGA furnace was measured at a $Y_2O_3$-$ZrO_2$ EMF cell placed in the gas line at the exit from the furnace. The specimen was then heated to, and held at, the desired temperature, at the known

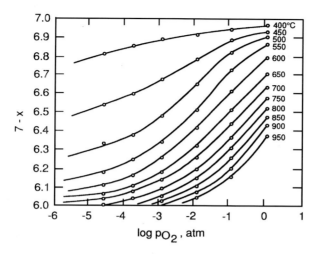

**Fig. 1** *The non-stoichiometry of $YBa_2Cu_3O_{7-x}$ as a function of temperature and the partial pressure of oxygen.*

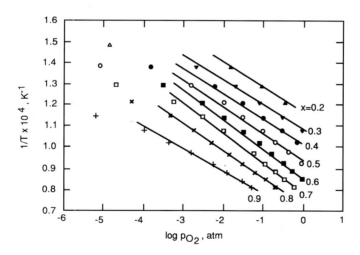

**Fig. 2** *Isocomposition lines in $YBa_2Cu_3O_{7-x}$. The dashed line is the state of the transition from the tetragonal phase to the orthorhombic phase.[4]*

oxygen pressure until it attained a constant weight. The temperature was then increased by 5°C and the specimen was maintained at the new constant temperature until it regained a constant weight. This procedure was repeated until measurements had been made over the required range of temperature. After a complete run the oxygen non-stoichiometry was determined by reducing the specimens in an atmosphere of Ar–4%H$_2$ at 920°C.

## Experimental Results

*The system YBa$_2$Cu$_3$O$_{7-x}$*

The dependence of oxygen non-stoichiometry of 123 on oxygen pressure and temperature is shown in Fig. 1 and the derived iso-composition curves are shown in Fig. 2. At oxygen pressures greater than $10^{-4}$ atm the iso-composition lines are linear which indicates that, in spite of the high value of $x$ in YBa$_2$Cu$_3$O$_{7-x}$, the system behaves as a solid solution which contains only one type of defect over a wide range of temperature and oxygen pressure. The results from the present study are compared with those of other studies in Fig. 3.

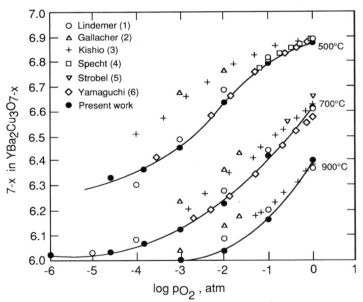

**Fig 3.** *A comparison of the results of the present study with those of previous studies. The measurements of Stobel et al.[6] were made at 685°C.*

The decomposition of YBa$_2$Cu$_3$O$_{7-x}$ was studied by thermogravimetric analysis and the products of the decomposition were identified by X-ray diffraction. Samples were held isothermally at fixed oxygen pressure and temperature until a constant weight was attained. The temperature was then increased by 5°C at a heating rate of 0.5°C/min and the sample was held isothermally until the new equilibrium state was reached. The oxygen pressure was varied in the range $10^{-5}$–$10^{-2}$ atm. The variation of $y$ in $(YBa_2Cu_3)O_y$ with temperature for a fixed oxygen pressure of $10^{-4}$ atm is shown in Fig. 4. The variation contains four plateaus over which the weight is independent of temperature, namely, Plateau I at $y \sim 6.0$, Plateau II at $y \sim 5.5$, Plateau III at $y \sim 5.3$ and Plateau IV at $y \sim 5.25$. Several specimens were held isothermally on Plateau I (820°C at $p_{O_2} = 10^{-4}$ atm and 900°C at $p_{O_2} = 10^{-3}$ atm) before

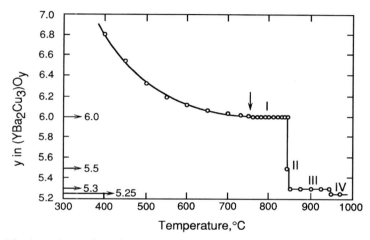

**Fig 4.** *The dependence of $y$ in $(YBa_2Cu_3)O_y$ on temperature at an oxygen pressure of $1.0 \times 10^{-4}$ atm.*

being quenched at a cooling rate of 500°C/min to room temperature. X-ray diffraction of the quenched specimens showed the presence of the phases 123 and 211 occurring in the ratio 92 : 8. No other phases were detected, which indicates that the decomposition reaction is very sluggish. In studies employing EMF cells Bormann and Nölting[8] and Porat *et al.*[10] reported that 123 decomposes to 211 + $BaCuO_2$ + $Cu_2O$. Thus, in view of the observation of no change in weight and the detection of 211, it can be deduced that the reaction

$$2YBa_2Cu_3O_6 \rightarrow Y_2BaCuO_5 + 3\ BaCuO_2 + Cu_2O \tag{1}$$

which does not involve a change in weight, occurs at the onset of Plateau I. The dependence, on oxygen pressure of the temperature at which the onset of Plateau I occurs is given by

$$\log_{10}p_{O_2}\ (\text{atm}) = \frac{-12\ 050}{T} + 7.64 \tag{2}$$

With further increase in temperature the oxygen content decreased abruptly to a value at which the nominal formula for the assemblage of phases occurring became $(YBa_2Cu_3)O_{5.3+\delta}$ ($\delta \leq 0.02$) on Plateau III. X-ray diffraction of samples quenched rapidly from Plateau III showed the presence of 211, $BaCu_2O_2$ and $YBa_3Cu_2O_x$, (132). However, the transformation of $211 + BaCuO_2 + Cu_2O$ to $211 + BaCu_2O_2 + 132$ violates the phase rule at the point of equilibrium by requiring that five condensed phases be in equilibrium with a gas phase in a four-component system at fixed temperature and oxygen pressure. This violation indicated that another three-phase field exists between Plateaus I and III. A more careful measurement of weight loss using 0.5–1°C increments in temperature showed another transformation at $T = 843°C$ with $p_{O_2} = 10^{-4}$ atm and at $T = 872°C$ with $p_{O_2} = 3.5 \times 10^{-4}$ atm which changes the nominal formula to $(YBa_2Cu_3)O_{5.5}$. This transformation occurs between Plateau I and Plateau III in Fig. 4. X-ray diffraction of samples quenched after being held on Plateau II for 15 days showed the presence of $BaCuO_2$ and $BaCu_2O_2$ which indicates that the transformation

$$BaCuO_2 + \tfrac{1}{2} Cu_2O \rightarrow BaCu_2O_2 + \tfrac{1}{4} O_2 \tag{3}$$

occurs between Plateau I and Plateau II, with the overall reaction being

$$2 YBa_2Cu_3O_6 \rightarrow Y_2BaCuO_5 + BaCuO_2 + 2 BaCu_2O_2 + \tfrac{1}{2} O_2 \tag{4}$$

Consequently, it is concluded that the transformation

$$Y_2BaCuO_5 + 7BaCuO_2 \rightarrow 2 YBa_3Cu_2O_{6+x} + 2 BaCu_2O_2 + \frac{(3-2x)}{2} O_2 \tag{5}$$

occurs between Plateau II and Plateau III, with the overall reaction being

$$7[(YBa_2Cu_3)O_6] \rightarrow 3 Y_2BaCuO_5 + YBa_3Cu_2O_{6+x} + 8 BaCu_2O_2 + \frac{(5-x)}{2} O2 \tag{6}$$

The dependence, on oxygen pressure, of the temperature at which the decomposition of $211 + BaCuO_2 + BaCu_2O_2$ to $211 + 132 + BaCu_2O_2$ is given by

$$\log_{10} p_{O_2} \text{(atm)} = \frac{-16\,980}{T} + 11.26 \tag{7}$$

With further increase in temperature the oxygen content of the sample decreased to Plateau IV in which state the nominal formula is $(YBa_2Cu_3)O_{5.24\pm0.01}$. X-ray diffraction of the green-coloured sample quenched from Plateau IV showed the presence of 211, $BaCu_2O_2$, $Ba(OH)_2 \cdot H_2O$ and $Ba(OH)_2$. Assuming that the $Ba(OH)_2 \cdot H_2O$ and $Ba(OH)_2$ were formed by reaction of the sample with moisture in the air during preparation of the specimen for X-ray diffraction, it is concluded that the transformation which takes the state of the system from Plateau III to Plateau IV is

$$4 YBa_3Cu_2O_{6+x} \rightarrow 2 Y_2BaCuO_5 + 3 BaCu_2O_2 + 7 BaO + \frac{(1+4x)}{2} O_2 \tag{8}$$

with the overall reaction being

$$2[(YBa_2Cu_3)O_6] \rightarrow Y_2BaCuO_5 + \frac{5}{2}BaCu_2O_2 + \frac{1}{2}BaO + \frac{3}{4}O_2 \qquad (9)$$

The stoichiometry of the right-hand side of Eq. (9), $(YBa_2Cu_3)O_{5.25}$ is close to the experimental observation of $(YBa_2Cu_3)O_{5.24}$.

The phase stability diagram constructed from the results of the present study is shown in Fig. 5.

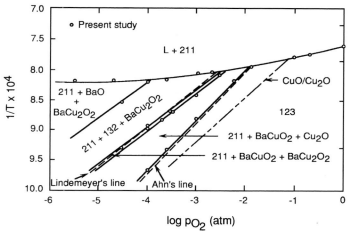

**Fig. 5** *The phase stability diagram for $(YBa_2Cu_3)O_y$.*

*The system $Nd_2CuO_{4-\delta}$*

The experimentally-measured variations of the apparent value of $4-\delta$ in $Nd_2CuO_{4-\delta}$ with oxygen pressure in the temperature range 700–950°C contained discontinuities at the temperatures at which the equilibrium

$$2CuO = Cu_2O + \frac{1}{2}O_2 \qquad (10)$$

occurs, which indicates that the specimens contained copper oxide as a separate phase. The magnitudes of the discontinuities gave 0.39 wt% as the proportion of CuO in the specimens, which is less than that required for detection by X-ray diffraction. Taking into consideration the weight decrease that occurs when the CuO impurity is reduced to $Cu_2O$, gives the oxygen non-stoichiometry of $Nd_2CuO_{4-\delta}$ as shown in Fig. 6. The maximum value of $4-\delta$ was determined to be 4.00.

The decomposition of $Nd_2CuO_{4-\delta}$ was studied using the same technique as used when studying the decomposition of 123 and the results obtained at three oxygen pressures are shown in Fig. 7. In each run, with increasing temperature, the value of $4-\delta$ in $Nd_2CuO_{4-\delta}$ decreases monotonically to the value of 3.93, at which temperature it abruptly decreases to the start of a plateau at $4-\delta = 3.50$. X-ray diffraction studies of specimens held at various temperatures and oxygen pressures

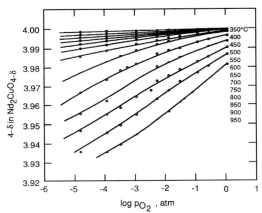

**Fig 6.** *The non-stoichiometry of $Nd_2CuO_{4-\delta}$ as a function of temperature and the partial pressure.*

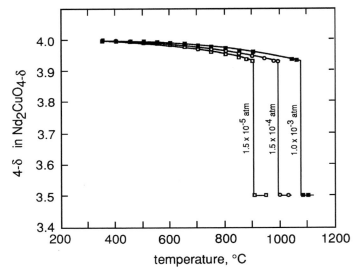

**Fig. 7** *The dependence of $4-\delta$ in $Nd_2CuO_{4-\delta}$ on temperature at oxygen partial pressures of $1.5 \times 10^{-5}$, $1.5 \times 10^{-4}$ and $1.0 \times 10^{-3}$ atm.*

for 40 h, before quenched into liquid nitrogen, indicated that the decomposition reaction is

$$4\,Nd_2CuO_4 \rightarrow 4\,NdCuO_2 + 2\,Nd_2O_3 + O_2 \tag{11}$$

and that, at slightly higher temperatures, the decomposition

$$2\,NdCuO_2 \rightarrow Nd_2O_3 + Cu_2O \tag{12}$$

28

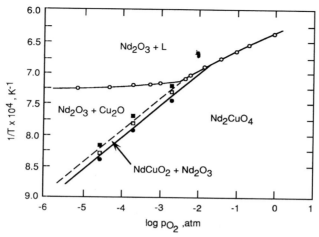

**Fig. 8**  *The phase stability diagram for $Nd_2CuO_4$.*

occurs. The transformation given by Eq. (12) does not involve a change in weight and hence cannot be detected by thermogravimetry. The deduced phase stability diagram, shown in Fig. 8, contains a narrow field of stability of $NdCuO_2 + Nd_2O_3$ between the fields of stability of $Nd_2CuO_4$ and $Nd_2O_3 + Cu_2O$. The line at which the equilibrium given by Eq. (11) occurs is

$$\log_{10} p_{O_2}\ (\text{atm}) = \frac{-18\,930}{T} + 11.26 \qquad (13)$$

and the position of the broken line which represents the equilibrium given by Eq. (12) was estimated from the results of the X-ray diffraction analyses.

*The system $Nd_{1.85}Ce_{0.15}CuO_{4-\delta}$*
The absolute oxygen content of $Nd_{1.85}Ce_{0.15}CuO_{4-\delta}$ was determined in the same manner as for $Nd_2CuO_{4-\delta}$ and the products of reduction in hydrogen were identified by X-ray diffraction as $Nd_2O_3$, Cu and $NdCeO_{3.5}$. The reduction of pure $CeO_2$ in hydrogen at 900°C produced the composition $CeO_{1.70}$, which indicates that the valence of some of the cerium is decreased to a value less than 4. However, in the reduction products of $Nd_{1.85}Ce_{0.15}CuO_{4-\delta}$, the cerium occurs in the compound $NdCeO_{3.5}$, which indicates that the valence of cerium is 4. Thus, assuming that the valences of Nd and Ce are not changed by reduction in hydrogen and that all of the copper is reduced to metallic Cu, the maximum value of $4-\delta$ in $Nd_{1.85}Ce_{0.15}CuO_{4-\delta}$ was calculated as 4.00.

The oxygen non-stoichiometry in $Nd_{1.85}Ce_{0.1}5CuO_{4-\delta}$ was also determined in the same manner as for $Nd_2CuO_{4-\delta}$. As was the case with $Nd_2CuO_{4-\delta}$, discontinuities were observed in the isotherms at oxygen pressures at which the $CuO–Cu_2O$ transformation occurs, and from the magnitudes of the apparent decreases in $4-\delta$ occurring at the discontinuities it was calculated that the specimen contained

29

1.03 wt% CuO, present as a separate phase. The presence of this CuO was taken into consideration in calculating the oxygen non-stoichiometry in $Nd_{1.85}Ce_{0.85}CuO_{4-\delta}$ shown in Fig. 9. Scanning electron microscopy and electron beam microprobe analysis of the prepared specimens of $Nd_{1.85}Ce_{0.85}CuO_{4-\delta}$ showed significant variation in homogeneity, in spite of the careful preparation procedures used. The standard deviation of measurements of the atomic concentration of Ce was 4.5% in comparison with 0.6% for Nd and 0.5% for Cu. This variation in the concentration of Ce indicates that $CeO_2$ is unusually unreactive during solid-state sintering.

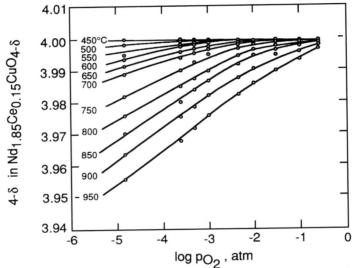

**Fig. 9**  *The non-stoichiometry of $Nd_{1.85}Ce_{0.15}CuO_{4-\delta}$ as a function of temperature and the partial pressure of oxygen.*

The phase stability diagram for $Nd_{1.85}Ce_{0.15}CuO_{4-\delta}$ was determined in the same manner as for $Nd_2CuO_{4-\delta}$ and the product of complete decomposition in the solid state had the nominal composition $Nd_{1.85}Ce_{0.15}CuO_{3.57}$. Samples were held for 57 h at various temperatures and oxygen pressures, before being quenched in liquid nitrogen and subjected to X-ray diffraction analysis. From the nominal composition and the observation of $Nd_2O_3$, $NdCeO_{3.5}$ and $Cu_2O$ as the products of composition, it is deduced that the decomposition reaction is

$$20\ Nd_{1.85}Ce_{0.15}CuO_{4-\delta} \rightarrow 3\ NdCeO_{3.5} + 17\ Nd_2O_3 + 10\ Cu_2O + (4.25-10\delta)\ O_2 \quad (14)$$

which occurs at

$$\log_{10}p_{O_2} = \frac{-19\,520}{T} + 10.95 \quad (15)$$

The stability diagram is shown in Fig. 10.

30

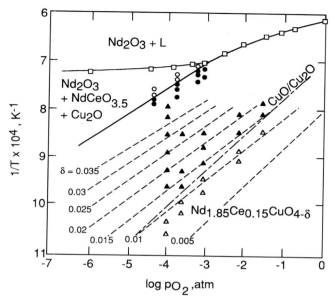

**Fig. 10**  *The phase stability diagram for $Nd_{1.85}Ce_{0.15}CuO_{4-\delta}$*

## Discussion and Conclusions

*The system $YBa_2Cu_3O_{7-x}$*

The results of the present study, together with those of previous studies, are shown in Fig. 11, which shows the non-stoichiometry, and structural and thermodynamic stability of the 123 system of stoichiometric composition ratio Y : Ba : Cu = 1 : 2 : 3.

Line 1 in Fig. 11 is for the tetragonal/orthohombic phase transition determined by Specht *et al.*[4] The temperature at which this transition occurs is significantly dependent on oxygen pressure. Line 2 is the decomposition of 123 to 211 + $Ba_2Cu_3O_{6-\delta}$ + CuO at high oxygen pressures ($p_{O_2} \geq 1.0$ atm), determined by Williams *et al.*,[18] and shows that 123 can be decomposed by oxidation as well as by reduction.

The iso-composition lines of 123 are shown as dotted lines. The 123 phase begins to decompose along line 4, which is also the iso-composition line of $x = 6.0$ in $YBa_2Cu_3O_x$. Using a solid-state electrochemical method, Nowotny *et al.*[19] identified $x = 6.0$ at $T = 800°C$ as a decomposition point for $YBa_2Cu_3O_x$, and Bormann and Nölting[8] reported $x = 6.1$ at $640°C \leq T \leq 750°C$. Porat *et al.*[10] reported that $YBa_2Cu_3O_x$ decomposes to other oxides when the value of $x$ is decreased to less than 6.1, and that the decomposition is reversible when the value of $x$ is greater than 5.97.

In the present study only the 211 phase was observed as a product of the decomposition of 123 in states above line 4. Bormann and Nölting[8], Ahn *et al.*[9] and Porat *et al.*[10] reported that 123 decomposes to 211 + $BaCuO_2$ + $Cu_2O$ and Lindemer *et al.*[7] and Ahn *et al*, in their later paper[20] have claimed that 123 decomposes to 211 + 132 + $BaCu_2O_2$. The ambiguity as to the nature of the decomposition products is

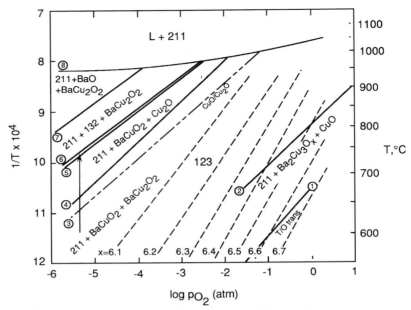

**Fig. 11** *The comprehensive phase stability diagram for YBa$_2$Cu$_3$O7$_{-x}$.*

probably caused by the extremely slow mobility of cations and/or lattice translations required to form new phases from the 123 matrix. In the present study the 211 phase was detected as a produce of decomposition only on the surface of the sample.

Line 5 is the boundary between the fields of stability of 211 + BaCuO$_2$ + Cu$_2$O and 211 + BaCuO$_2$ + BaCu$_2$O$_2$. This line is less than 5°C from line 6, which is the boundary between the fields of stability of 211 + BaCuO$_2$ + BaCu$_2$O$_2$ and 211 + 132 + BaCu$_2$O$_2$. Line 6 was incorrectly identified by Lindemer et al.[7] as being the line at which 123 decomposes to 211 + 132 + BaCu$_2$O$_2$. Line 7 is the boundary between the fields of stability of 211 + 132 + BaCu$_2$O$_2$ and 211 + BaCu$_2$O$_2$ + BaO and line 8 is the solidus line determined by differential thermal analysis.

Considering the quaternary phase diagram for the system Y$_2$O$_3$–BaO–Cu–O at a fixed temperature and fixed stiochiometry of Y : Ba : Cu = 1 : 2 : 3, Fig. 11 shows that decreasing the oxygen pressure causes the state of the system to pass through four stability fields in which three condensed phases are in equilibrium with the gas phase. For example, at 900°C, (1/$T$ = 8.5 x 10$^{-4}$), decreasing the oxygen pressure causes 123 to decompose to 211 + BaCuO$_2$ + Cu$_2$O at $p_{O_2}$ = 1.5 x 10$^{-3}$ atm, and then causes transitions to 211 + BaCuO$_2$ + BaCu$_2$O$_2$ at $p_{O_2}$ = 2.9 x 10$^{-4}$ atm, to 211 + 132 + BaCu$_2$O$_2$ at $p_{O_2}$ = 2.3 x 10$^{-4}$ atm and to 211 + BaCu$_2$O$_2$ + BaO at $p_{O_2}$ = 1.4 x 10$^{-5}$ atm.

In Fig. 10 the filled triangles are thermodynamic states of processing which produce superconductivity in the oxide, the open triangles are states of processing which do not give a superconducting oxide and the dash-dot-dash line is the variation, with temperature, of the oxygen pressure at which the equilibrium

$$Cu_2O + \tfrac{1}{2}O_2 = 2\,CuO$$

occurs in the binary system Cu–O.[17] As is seen, this line occurs at the boundary between the processing field which gives a superconducting oxide and the processing field which gives a non-superconducting oxide. The reason why superconductivity is obtained by processing in thermodynamic states in which $Cu_2O$ is stable relative to $CuO$, and is not obtained by processing in states in which $CuO$ is stable relative to $Cu_2O$ is not apparent.

The variances of oxygen non-stiochiometry in $Nd_2CuO_{4-\delta}$ and $Nd_{1.85}Ce_{0.15}CuO_{4-\delta}$, which are, respectively, $\delta = 0.0 - 0.07$ and $\delta = 0.0 - 0.045$, in the temperature range $350°C - 950°C$ and the range of oxygen pressure $10^{-5}$ - 1 atm are very small in comparison with variance of $x = 0.0 - 1.0$ which occurs in the system $YBa_2Cu_3O_{7-x}$, (123). The phase stability diagrams for the two systems show that, with decreasing oxygen pressure, $Nd_2CuO_{4-\delta}$ decomposes to $NdCuO_2 + Nd_2O_3$, which, with further decrease in oxygen pressure, decompose to $Nd_2O_3 + Cu_2O$. $Nd_{1.85}Ce_{0.15}CuO_{4-\delta}$ decomposes to $Nd_2O_3 + NdCeO_{3.5} + Cu_2O$. The stability field of $Nd_{1.85}Ce_{0.15}CuO_{4-\delta}$ contains a sub-field in which processing produces a superconducting oxide and a sub-field in which processing produces an oxide which is not superconducting. These two sub-fields meet at the line along which $Cu_2O$ and $CuO$ coexist in equilibrium in the system Cu-O.

# References

1. T. B. Lindemer, J. F. Hunley, J. E. Gates, A. L. Sutton, Jr., J. Brynestad, C. R. Hubbard and P. K. Gallagher: 'Experimental and Thermodynamic Study of Nonstoichiometry in $YBa_2Cu_3O_{7-x}$,' *J. Am. Ceram. Soc.*, 1989, **72** (10), 1775–1788.
2. P. K. Gallagher: 'Characterization of $Ba_2YCu_3O_x$ as a Function of Oxygen Partial Pressure Part I: Thermoanalytical Measurements,' *Adv. Ceram. Mater.*, 1987, **2** (3B), 632–639.
3. K. Kishio, J. Shimoyama, T. Hasegawa, K. Kitazawa and K. Fueki: 'Determination of Oxygen Nonstoichiometry in a High-$Tc$ Superconductor $YBa_2Cu_3O_{7-x}$,' *Jpn. J. Appl. Phys.*, 1987, **26**, L1228–1230.
4. E. D. Specht, C. J. Sparks, A. G. Dhere, J. Brynestard, O. B. Cavin, D. M. Kroeger and H. A. Oye: 'Effect of Oxygen Pressure on the Orthohombic-Tetragonal Transition in the High Temperature Superconductor $YBa_2Cu_3O_x$,' *Phys. Rev. B: Condens. Matter*, 1988, **37**, 7426–7434.
5. P. Strobel, J. J. Capponi and M. Marezio: 'High Temperature Oxygen Defect Equilibrium in Superconducting Oxide $YBa_2Cu_3O_{7-x}$,' *Solid State Commun.*, 1987, **64**, 513–515.
6. S. Yamaguchi, K. Terabe, A. Saito, S. Yahagi and Y. Iguchi: 'Determination of Nonstoichiometry in $YBa_2Cu_3O_{7-x}$,' *Jpn. J. Appl. Phys.*, 1988, **27**, L179–181.

7. T. B. Lindemer, F. A. Washburn, C. S.MacDougall and O. B. Cavin: 'Synthesis of Y–Ba–Cu–O Superconductors in Subatmospheric Oxygen,' *Physica C*, 1991, **174**, 135–143.

8. R. Bormann and J. Nölting, 'Stability Limit of the Perovskite Structure in the Y–Ba–Cu–O Systems,' *Appl, Phys. Lett.*, 1989, **54**, (21), 2148–2150.

9. B. T. Ahn, T. M. Gür, R. A. Huggins, R. Beyers, E. M. Engler, P. M. Grant, S. S. P. Parkin, G. Lim, M. L. Ramirez, K. P. Roche, J. E. Vazquez, V. Y. Lee, and R. D. Jacowitz: 'Studies of Superconducting Oxides with a Solid-state Ionic Technique,' *Physica C*, 1988, **153-155** 590–593.

10. O. Porat, I. Reiss, and H. L. Tuller: 'Investigation of Defect Equilibrium in $YBa_2Cu_3O_x$ by a Solid State Electrochemical Method,' *Physica C*, 1992, **192**, 60–74.

11. C. N. Pieczulewski, K. S. Kirkpatrick and T. O. Mason: '1000°C Phase Relationships and Superconductivity in th (Nd–Ce–Cu)–O Ternary System ' *J. Am. Ceram. Soc.*, 1990, **73**, 2141.

12. K. Oka and H. Unoki: 'Phase Diagrams and Crystal Growth of Superconductive $(NdCe)_2CuO_4$' *Jpn. J. Appl. Phys.*, 1989, **28**, L937.

13. E. Moran, A. I. Nazzal, T. C. Huang and J. B. Torrance: 'Extra Oxygen in Electron Superconductors: Ce and Th Doped $Nd_2CuO_{4+\delta}$ and $Gd_2CuO_{4+\delta}$' *Physica C*, 1989, **160**, 30.

14. E. Wang, J. M. Tarascon, L. H. Greene and G. W. Hull: 'Cationic Substitution and the Role of Oxygen in the *n*-type Superconducting $T'$ System $Nd_{2-y}Ce_yCuO_x$,' *Phys. Rev. B.*, 1990, **41**, 6582.

15. K. Suzuki, K. Kishio, T. Hasagawa and K. Kitazawa: 'Oxygen Nonstoichiometry of the $(Nd,Ce)_2CuO_{4-\delta}$ System,' *Physica C.*, 1990, **166**, 357.

16. A. N. Petrov, V. A. Cherepanov, A. Yu Zuyer and V. M. Zhukovsky: 'Thermodynamic Stability of Ternary Oxides in the Ln–M–O (Ln = La, Pr, Nd; M = Co, Ni, Cu) Systems,' *J. Solid State Chem.*, 1988, **77**, 1.

17. R. Schmid: 'A Thermodynamic Analysis of the Cu–O System with an Associated Solution Model,' *Met. Trans. B*, 1983, **14**, 473–481.

18. R. K. Williams, K. B. Alexande, J. Brynestad, T. J. Henson, D. M. Kroeger, T. B. Lindemer, G. C. Marsh and E. D. Specht: 'Oxidation-Induced Decomposition of $YBa_2Cu_3O_{7-x}$,' *J. Appl. Phys.*, 1991, **70** (2), 906–913.

19. J. Nowotny, M. Rekas and W. Weppner, 'Defect Equilibria and Transport in $YBa_2Cu_3O_{7-x}$ at Elevated Temperatures: I Thermopower, Electrical Conductivity, and Galvanic Cell Studies,' *J. Am. Ceram. Soc.*, 1990, **73** (4), 1040–1047.

20. B. T. Ahn, V. Y. Lee, R. Beyers, T. M. Gür, and R. A. Huggins, 'High Temperature Phase Equilibria near $YBa_2Cu_3O_{6+x}$,' *Physica C*, 1989, **162–164**, 883–884.

# 'T-Jump' Experiments for Determining High Temperature Transport Properties of YBCO

G. Balducci, F. Cellucci, D. Gozzi*

*Dipartimento di Chimica, Università 'La Sapienza' Ple., Aldo Moro 5, 00185 Rome, Italy*

AND M. Tomellini

*Dipartimento di Scienze e Tecnologie Chimiche, Università di 'Tor Vergata' Via della Ricerca Scientifica, 00133, Rome, Italy*

## Abstract

Oxygen diffusion in YBCO has been investigated by means of a non-equilibrium T-jump experiment. The kinetic curves have been analysed in the temperature range 600-750°C on the basis of a bipolar transport model leading to the evaluation of the activation energy for oxygen vacancy formation and diffusion.

The ionic transport properties of the $YBa_2Cu_3O_{7-\delta}$ (YBCO) mixed conductor have been widely studied by means of either thermogravimetric techniques or direct measurements of ionic conductivity.[1,2] Thermodynamic and kinetic quantities, namely the enthalpy change occurring during the defect formation and the activation enthalpy for oxygen diffusion, have been determined for oxygen transport in both orthorhombic and tetragonal phases of YBCO and at different values of the oxygen fugacity.[2] Oxygen transport in orthorhombic YBCO has also been studied through a thermogravimetric technique by means of a non-equilibrium 'P-jump' experiment,[3,4] in which the weight loss or the resistivity of the sample is continuously recorded after the rapid decrease of the oxygen partial pressure in the gas phase. The kinetics data have been processed according to the Fick's second law and an activation enthalpy of about 1.3 eV for oxygen diffusion in YBCO orthorhombic phase was found.[3]

In our previous paper we have presented a kinetic study for YBCO–$O_2$ heterogeneous equilibrium by a non-equilibrium 'T-Jump' experiment.[5] At a fixed value of the oxygen fugacity the superconductor oxide (sample sintered at $T=950°C$ with density equal to 6 g cm$^{-3}$) is submitted to a sudden change of the temperature producing a time-dependent concentration of the vacancy defects in the solid phase. The relaxation process between the two equilibrium states is studied through the measurements of the time-dependence of the electronic resistivity of the sample, $\rho$, by means of the four-probe technique. The average value of the oxygen vacancy fraction, $\delta$, has been linked to the electronic resistivity on the ground of Fiory's expression.[5,6]

As an example in Fig. 1 the $\delta(t)$ function is reported for a temperature jump of 50°C

* To whom correspondence should be addressed.

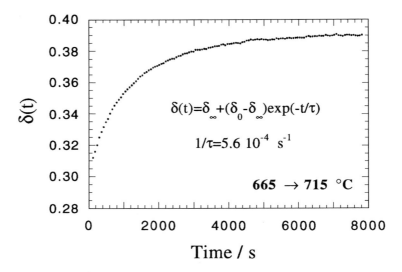

**Fig. 1** *Typical trend of the defectivity, δ, as a function of the time, t, after a 'T-jump' experiment. The relaxation curve is well described by a first order kinetic law, in agreement within a kinetic model developed on the basis of the kinetic law of mass action.*

($665 \rightarrow 715$°C) and at an oxygen partial pressure of 1 atm. These data are described by a first order kinetic law in agreement with the kinetic model previously developed on the basis of the *kinetic law of mass action*.[5] In particular, two heterogeneous reactions were proposed that account for the formation of both single and double charged oxygen vacancies:

$$M'_M + V_O^\bullet + 1/2\ O_2 = M_M + O_O$$

and

$$2M'_M + V_O^{\bullet\bullet} + 1/2\ O_2 = 2M_M + O_O$$

where $M$ stands for a metal ion, $O_O$ for an oxygen atom in its regular position in the basal plane and $V_O^{\bullet\bullet}$ ($V_O^\bullet$) for a double (single) positive charged oxygen vacancy. In these reactions it is assumed that the excess charges are localised and, therefore, assigned to one of the species. In other words the $M'_M$ species plays the role of a donor state where the charge transfer leaves a hole, namely $M_M$. The equilibrium constants of the heterogeneous reactions for the formation of single and double charged oxygen vacancies are given by:[5]

$$K_1 = \frac{(2-\delta)\,(1-\delta)}{(P_{O_2})^{1/2}\,\delta^2} \tag{1}$$

$$K_2 = \frac{(1-\delta)^3}{(P_{O_2})^{1/2}\,\delta^3} \tag{2}$$

36

where $\delta$ is the oxygen defectivity, $K_{1(2)}$ are the equilibrium constants for single and double charged vacancies and $P_{O_2}$ is the oxygen partial pressure. To identify the solid-state process involved during oxygen exchange the dependence of $\rho$ on the oxygen pressure was analysed. The experimental data seem to support an oxygen exchange mechanism based on the formation of single charged vacancies. However, an analysis of the ionic transport properties of YBCO in the light of the non-equilibrium T-jump experiment is still lacking. The purpose of this work is therefore to derive thermodynamic and kinetic quantities for the oxygen transport in YBCO, by processing of the data mentioned above.

Let us now focus our attention to the YBCO system as a *mixed* conductor in which the electronic conductivity is higher than the ionic conductivity. Because of the gradient of the oxygen chemical potential across the sample during the non-equilibrium experiments, a flux of both vacancies, J, is given, according to the bipolar transport mechanism, as follows:[7]

$$J_i = - \frac{1}{z^2 F^2} \frac{\sigma_e \sigma_v}{\sigma_e + \sigma_v} \left( \frac{d\mu_v}{d\xi_i} \right) \tag{3}$$

where $\xi_i$ is the spatial coordinate along the $i$ direction, $\sigma_e$, $\sigma_v$ are the electronic and ionic conductivities, $z$ is the vacancy charge, $F$ the Faraday's constant and $\mu_v$ the chemical potential of *neutral* oxygen vacancies that is of the form: $d\mu_v = RT d\ln C_{V_O}$, $C_{V_O}$ being the concentration of vacancies. We now link the concentration of neutral vacancies to the one of charged vacancies on the basis of the equilibrium $zM'_M + V_O{}^{z+} = zM_M + V_O$ providing $C_{V_O} \propto (C_{V_O{}^{z+}})^{z+1}$ with $z = 1, 2$.[5] On the ground of the Nernst Einstein equation and considering $\sigma_v \ll \sigma_e$, Eq. (3) reduces to:

$$J = -(z+1)D \nabla C_{V_O{}^{z+}} \tag{4}$$

and, through the continuity equation, to Fick's law in the form:

$$\frac{\partial C_{V_O{}^{z+}}}{\partial t} = (z+1) D \nabla^2 C_{V_O{}^{z+}} \tag{5}$$

where $D = D_0 \exp(-\Delta H_d / RT)$ is the diffusion coefficient of the charged vacancy, that is also a function of $z$.

Fick's law has been solved in ref. 3 for describing the YBCO weight change during a 'P-jump' experiment. In that model each particle of the sample is assumed to be a circular stab of aribitrary thickness and radius equal to $R$. The solution provides a first order kinetic law with a time constant given by $R^2/(\lambda D)$ where $\lambda$ is a constant.[3] If the solution in the literature[3] is now used for Eq. (5), one can easily calculate the time constant of a 'T-jump' experiment by the expression $\tau = R^2/[\lambda(z+1)D]$. It is important to point out that the transport process studied on the ground of Wagner's theory leads to a transport equation that depends on the vacancy charge and, consequently, the time constant is a function of $z$.

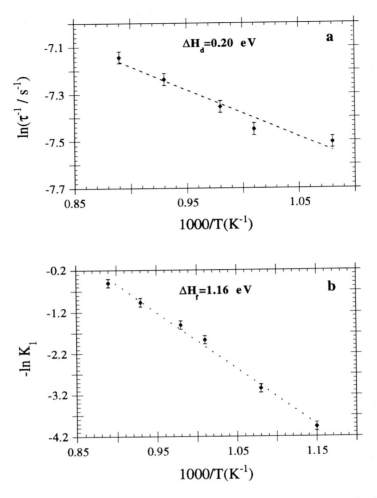

**Fig. 2** *Arrhenius plot of $\tau^{-1}$ leading to the estimation of the activation energy for diffusion (panel a). In panel b the Arrhenius plot of the equilibrium constant $K_1$ (Eq. [1] ) is also reported, providing for the evaluation of the formation enthalpy of oxygen vacancies.*

In Fig. 2a is reported the Arrhenius plot of the $\tau^{-1}$ quantity as obtained by the 'T-jump' experiments, and provides an activation energy for diffusion, $\Delta H_d$, equal to $0.20\pm0.05$ eV in agreement with values previously measured.[8] It is worth mentioning that the value of the diffusion coefficient is sensitive to the grain size, preferred orientation of grains, density and to the presence of any surface and grain boundary phase. In particular, the out-diffusion of oxygen was also found to be an interfacial-limited reaction with an activation energy of 1.7 eV in the temperature range from 300 to 440°C.[4] Our kinetic data seem to be representative of a diffusion limited process rather than of an interfacial limited process. In fact, an interfacial limited process should give a constant value of the outcoming flux of oxygen leading to a linear trend of the $\delta$ vs $t$ curve as observed in ref. 4. Moreover, if oxygen recombination at the YBCO surface was rate determining $[2O_{ad} \rightarrow O_2(gas)]$, one would still observe linear kinetics due to the constant value of the outcoming flux $j=K\theta_O^2$ K being the recombination constant and $\theta_O \approx 1$ the $O_{ad}$ surface coverage.

In Fig. 2b the $\ln(1/K_1)$ vs $1/T$ is also shown, which provides a standard activation energy for the vacancy formation, $\Delta H_f$, equal to $1.16 \pm 0.05$ eV, in agreement with the literature value.[2] Moreover, the activation energy for the ionic conduction is expressed in the form $\Delta H_i = (z+1)^{-1}\Delta H_f + \Delta H_d = 0.8$ eV for single charged oxygen vacancies.

Transport properties of the superconductor could also be derived by an appropriate analysis of the hysteresis cycles. At a fixed value of the oxygen pressure, the sample is submitted to a heating cycle at a constant heating rate and the resistivity continuously recorded. An example is shown in Fig. 3 where the hysteresis cycle of

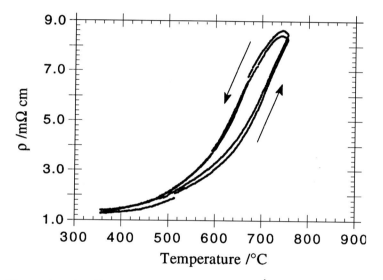

**Fig. 3** *Hysteresis cycle collected at a constant heating rate of $\dot{T} = 1\,°C\,min^{-1}$ and $P_{O_2}=1$ atm.*

the sample is reported for the heating rate of $dT/dt=1°C$ min$^{-1}$. At any time the resistivity of the specimens is a function of both temperature and defectivity, $\rho=\rho(T,\delta)$, where $T$ and $\delta$ are the actual values of $T$ and $\delta$. The total derivative of the resistivity is goven by:[9]

$$\frac{d\rho}{dt} = \left( \frac{\partial\rho}{\partial T} \right)_\delta \dot{T} + \left( \frac{\partial\rho}{\partial\delta} \right)_T \dot{\delta} \cong \alpha\dot{T}+\rho_o\dot{\delta} \qquad (6)$$

The $\dot{\delta}$ quantity is evaluated making use of Eq. (6) through the knowledge of $\alpha$, $\dot{T}$ and $\dot{\rho}$ on the basis of the hysteresis curves. As previously established,[5]

$\frac{d\delta}{dt} = (\delta_e - \delta)D > 0$ at $\dot{T} > 0$, where $\delta_e$ is the equilibrium value and the activation energy

for out diffusion can be estimated through the Arrhenius plot of the quantity $\frac{\dot{\delta}}{\Delta\dot{\delta}}$

The $\Delta\delta$ term was also found to be nearly constant in the explored temperature range (600–750°C). The result provides a $\Delta H_d$ value of 0.32 eV, that is in agreement with the one obtained from the T-jump experiments.

In conclusion, an attempt to determine the transport properties of YBCO has been proposed on the basis of non-equilibrium T-jump experiments. The experimental data were analysed according to a bipolar transport mechanism that is dependent on the charge of the ionic carriers, in contrast to Fick's second law which is usually invoked for explaining the oxygen vacancy transport in YBCO as *pure* diffusion processes.

## References

1.  V. Carrillo-Cabrera, H-D. Wiemhöfer and W. Göpel: *Solid State Ionics*, 1989, **32/33**, 1172.
    J. L. MacManus, Derek J. Fray and Jan E. Evatts: *Physica C*, 1992, **190**, 511.
2.  P. Strobel, J. J. Capponi, M. Marezio and P. Monod: *Solid State Comm.*, 1987, **64**, 513.
3.  T. B. Tang and W. Lo: *Physica C*, 1991, **174**, 463.
4.  K. N. TU, S. I. Park and C. C. Tsuei: *App. Phys. Lett.*, 1987, **51**, 2158.
5.  F. Cellucci, P. L. Cignini, D. Gozzi and M. Tomellini: *J. Mater. Chem.*, 1993, **3**, 421.
6.  A. T. Fiory, M. Gurvitch, R. J. Cava and G. P. Espinosa: *Phys. Rev. B*, 1987, **36**, 7262.
7.  Per Kofstad: *High Temperature Oxidation of Metals*, John Wiley & Sons, Inc. New York, 1966.
8.  E. Size and J. S. Moya: *Mat. Lett.*, 1988, **6**, 369.
9.  F. Cellucci, D. Gozzi and M. Tomellini: *J. Mater. Chem.*, issue 4 1994.

# The Phase Transformation in $\beta$-$(Bi_2O_3)_{1-x}(SrO)_x$

H. W. KING AND E. A. PAYZANT

*Materials Engineering, University of Western Ontario, London, ON, N6A 5B9, Canada*


C. B. ALCOCK AND Y. SHEN

*Center for Sensor Materials, University of Notre Dame, South Bend, IN 46556, USA*

### Abstract

Significant differences occur in the location of phase boundaries of solid solutions in ceramic oxide systems that are susceptible to displacive phase transformations. In many cases the form of the published diagram is a direct function of the method of investigation. A case in point is the $\beta$-phase in the $Bi_2O_3$–SrO system, which has attracted attention as a potential solid oxide electrolyte. On the basis of differential thermal analysis and high temperature X-ray diffraction, some investigators report an order : disorder transition based on a rearrangement of oxide ion vacancies, while others report no evidence of such a transition using optical microscopy and X-ray diffraction of quenched samples. Careful *in situ* X-ray diffraction measurements show that the lattice parameters of the $\beta$-phase increase sharply at high temperatures and that the temperature of this discontinuity is composition dependent, being a maximum at compositions at the centre of the phase. This effect is interpreted in terms of a sudden relaxation of the M—O bonds within the crystal structure of the $\beta$-phase, resulting in increased oxide ion conductivity at high temperatures.

## Introduction

When Professor Alcock was a student some 50 years ago, science was in a confident mood and a great emphasis was placed on careful experimentation. At that time it was common to speak of 'equilibrium diagrams', even though everyone knew that the underlying theoretical condition was never really achieved in practice. When science entered a more cautious phase some 20 years later, the nomenclature changed to 'phase diagrams' and the emphasis shifted from pure experiment to a combination of experiment and prediction based on thermodynamic data. Since the effectiveness of the latter approach is a direct function of the quality of the available data, it is most appropriate at this time to acknowledge the key contribution of Professor Alcock, who by careful collection, appraisal and storage has built up an internationally recognised series of reliable and accessible thermodynamic data banks.

The combination of theory and experiment does not always produce the required phase diagram, however, due to the lack of diffusion in the solid state at low temperatures and the concurrent operation of diffusionless phase transformations

41

that produce metastable phases without a change in composition. Although the latter have been known for many years, it is conveniently assumed that they are limited to one or two classical systems and can thus be ignored when investigating the rest. In the classical metal system Fe–Ni, the diagram describing the temperature–composition dependence of phases generated by diffusionless martensitic transformations is pragmatically referred to as a 'realisation diagram',[1] while the 'phase diagram' itself is reserved for the investigation of very slowly cooled alloys, such as metallic meteorites. In the classical oxide systems, silica transforms on slow cooling from crystobalite → tridymite → quartz, by a diffusion controlled process that is descriptively referred to as 'reconstructive'.[2] However, rapid cooling of any of these structures induces a diffusionless transformation to a 'middle' or 'lower' form, by a so-called 'displacive' transformation.[2] These transformations can lead to significant differences between phase boundaries derived from quenched versus slowly cooled samples, as observed for the partially stabilised zirconias, where the temperature determined for the $\gamma \rightarrow \alpha$ transformation can vary by as much as 200°C.[3]

The structural stability of the β-phase of the oxide system $Bi_2O_3$–SrO is also the subject of some confusion. This phase is of potential interest for fuel cell applications, because its oxide ion conductivity in oxidising atmospheres is significantly greater than that of the stabilised zirconias and doped perovskites.[4] The ionic conductivity increases by an order of magnitude at 620–670°C and maintains an exceptional high value up to 800°C.[4] The crystal structure of the β-phase was investigated by Sillén and Aurivillius,[5, 6] who defined a defect rhombohedral structure in which the cation sites in the basal plane are occupied solely by Bi ions, while sites in intermediate layers are randomly occupied by both Bi and Sr ions. Three types of oxide ion sites were also identified, but in order to maintain the electrical neutrality of the crystal only a fraction of these are considered to be occupied, with the resultant oxygen vacancies distributed at random among the different sites. The high ionic conductivity of the β-phase has been attributed to the number and mobility of these oxide ion vacancies.[4]

As shown in the $Bi_2O_3$–SrO phase diagram in Fig. 1, the β-phase is an intermediate solid solution which extends from 17.5–42.5 mol.% SrO, but conflicting maximum composition boundaries are indicated at high temperatures. Using differential thermal analysis and high temperature X-ray diffraction, Guillermo et al. have proposed that the solution of SrO in the β-phase increases from 42.5 mol.% at 790°C to 55 mol.% at 930°C, when the phase decomposes incongruently.[7] By contrast, on the basis of optical microscopy and X-ray diffraction using quenched samples, Hwang et al. have reported that the composition range of the β-phase remains unchanged on heating to 925°C, and then narrows to a congruently melting composition of 41 mol.% SrO at 960°C.[8] Guillermo et al. also propose a phase transition in the β-phase at temperatures between 680°C and 715°C, as evidenced by a discontinuous expansion of the $c$ parameter of the crystal lattice,[7] while Hwang et al. report that no evidence of this transition was observed in their quenched samples.[8] In an attempt to clarify the situation with respect to the proposed transition in the β-phase, and incidentally provide a rationale of the marked increase observed

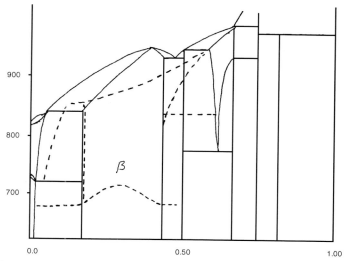

**Fig. 1** *Phase diagram of the $Bi_2O_3$–SrO system, showing the β-phase boundaries proposed by Guillermo et al.[7] and Hwang et al.[8]*

in oxide ion conductivity at 620–670°C, samples of the $\beta-(Bi_2O_3)_{1-x}(SrO)_x$ phase with various compositions have been prepared and examined *in situ* by high temperature X-ray diffraction.

## Experimental

Proportionate amounts of $Bi_2O_3$ (99.9%) and $Sr(NO_3)_2$ (99.9%) were reacted in the solid state to produce 20 g samples of $(Bi_2O_3)_{1-x}(SrO)_x$ with compositions $x = 0.20$, 0.32 and 0.40. The powders were mixed and ground and then calcined in air for 10 h at 800°C. The reacted powders were uniaxially pressed to produce discs 25 mm in diameter and ~5 mm in thickness, which were sintered for 10 h in air at 800°C. After sintering, the disks were crushed, ground, pressed and fired two additional times, to obtain improved homogeneity.

The electrical conductivity of the sintered disks was measured with an AC impedance spectrometer at a frequency of 7 mHz to 100 mHz. Using sputtered Au–Pd electrical contacts, the conductivity of the disk samples was measured in air over the temperature range from 400°C to 800°C by heating the samples in a tube furnace, the temperature of which was measured and controlled to ± 2°C.

The crystal structures of powdered samples were examined at room temperature, and at 400–800°C, using a high temperature attachment mounted on a computer-controlled theta : theta X-ray diffractometer, with copper Kα radiation. For structural identification and lattice parameter determination, diffraction profiles were step-scanned at an equivalent scanning speed of 1 °/min., using a step interval of 0.03° and a 0.3 mm receiving slit. X-ray diffraction patterns were determined in air, using a Pt/Rh heater strip which enabled the temperature to be set and controlled to ±

1°C during heating and cooling cycles. At least 27 peaks were recorded at each temperature, for lattice parameter determinations. The background and $K\alpha_2$ components were removed by a background substraction program and the positions of the diffraction peaks were then determined with a profile fitting program based on the Pearson VII function.[9] The rhombohedral structure was indexed with respect to hexagonal indices and the lattice parameters were derived by a computational method equivalent to that of Vogel and Kempter,[10] based on the error function $\cos\theta\cot\theta$,[11] and also by a least squares method developed by Burnham.[12]

## Results and Discussion

As shown by the electrical conductivity results for the $\beta$-$(Bi_2O_3)_{1-x}(SrO)_x$ phase samples plotted in Fig. 2, at temperatures up to ~550°C the magnitude of the observed conductivities decreases in the order of increasing composition, i.e. 0.20 > 0.32 > 0.40. The conductivity of all of the samples increases sharply in the temperature range 600–700°C, but since the onset of this effect is a function of composition, the plots crossover in this temperature region. At higher temperatures the magnitude of the conductivity is again in the order of increasing composition, i.e. 0.20 > 0.32 > 0.40. This form of temperature and composition dependence for the conductivity of the $\beta$-$(Bi_2O_3)_{1-x}(SrO)_x$ phase is similar to that observed previously by Takahashi et al.[4]

The $\beta$-$(Bi_2O_3)_{1-x}(SrO)_x$ phase samples were bright yellow in appearance in the as-

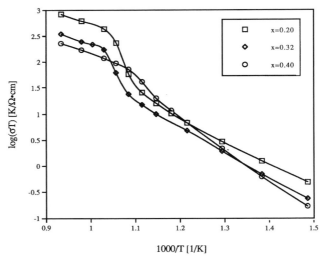

**Fig. 2** *High temperature conductivities of the $\beta$-$(Bi_2O_3)_{1-x}(SrO)_x$ phase in air.*

prepared condition and X-ray examination showed that all of the samples have the rhombohedral structure identified by Sillèn and Aurivillius.[5,6] However, after storing for up to three months at room temperature, the appearance of the samples changed to a pale dull yellow and X-ray examination diffraction patterns included peaks of

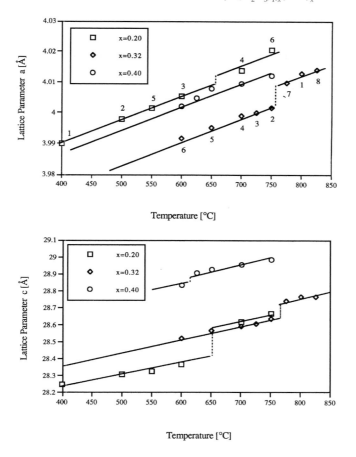

**Fig. 3** *Temperature dependence of lattice parameters of the β-(Bi₂O₃)₁₋ₓ(SrO)ₓ phase in air.*

the Bi₂O₃ structure, indicating that the β-phase is susceptible to decomposition in a mildly damp atmosphere at room temperature. After the hydrated samples were heated for 24 h at 700°C, the bright yellow colour was restored and the Bi₂O₃ peaks were no longer observed in the diffraction patterns.

Lattice parameters derived from *in situ* X-ray diffraction measurements over the temperature range 400–800°C were calculated in terms of the hexagonal unit cell of the rhombohedral β-phase structure. Sharp discontinuities were observed in both the *a* and *c* lattice parameters in the temperature range 650–750°C, as shown in Fig. 3, but no associated changes were observed in the relative intensities of the 00.*l* and *hh*.0 diffraction peaks. To confirm the reproducibility of these lattice parameter results, sequential X-ray diffraction measurements were made above and below the temperatures of the respective discontinuities, as indicated by the numerical sequence of the data points plotted in Fig. 3. The reliability of the computer methods

45

used to derive the lattice parameters of the hexagonal unit cell is also indicated by the lack of scatter between the results obtained by applying both the extrapolation and least squares programs to the same set of experimental measurements.

The lattice parameter results plotted in Fig. 3 show that the temperature of the sharp discontinuity is composition dependent, occurring at ~650°C for $x = 0.20$, ~750°C for $x = 0.32$ and at ~625°C for $x = 0.40$. These temperatures coincide with the temperatures of the sharp increases in conductivity for the samples with $x = 0.20$ and 0.40, but the temperature of the lattice parameter discontinuity in the sample with $x = 0.32$ is about 100°C higher than the temperature for the increase in conductivity. Even so, both of these effects occur at a maximum temperature for compositions in the centre of the phase, in confirmation of the findings reported previously by Guillermo et al.[7] These authors focused their attention on the sharp increase in $c$ parameter, which was interpreted in terms of some rearrangement within the lattice of oxygen vacancies. However, since the abrupt increase occurs in both the $a$ and $c$ parameters, it is more indicative of a sudden relaxation of the M—O bonds throughout the structure, resulting in a discontinuous increase in volume. Such a relaxation would also be consistent with the absence of any change in the relative intensities of the 00.$l$ and $hh$.0 diffraction peaks, which would be indicative of an order : disorder transition. While further study is necessary to determine the details of the relaxation of the M—O bonds, the onset of such an effect would greatly increase the diffusion of oxygen vacancies in all directions in the crystal structure of the β-phase, resulting in the observed increases in oxide ion conductivity at high temperatures.

The present study confirms the necessity of applying a range of investigative techniques to the investigation of phase diagrams of oxide systems.[13] While the use of quenched samples may be satisfactory for determining the shape of the liquidus, as shown by Hwang et al.,[8] this method is quite inappropriate for detecting displacive transitions. The various experimental methods are also to be regarded as collaborative, rather than competitive, and minor anomalies in physical properties, or small peaks in differential thermal analysis, must all be integrated into a diagram which is consistent with thermodynamic principles. Further, since one person cannot be a master all the techniques, it is productive for scientists from different backgrounds to cooperate in the investigation of complex systems. In this regard, it is appropriate to finish by acknowledging that the present contribution to the understanding of the transition in the β-$(Bi_2O_3)_{1-x}(SrO)_x$ phase has only been accomplished by combining the experience in physical techniques with the chemical expertise of Professor Alcock, whom we honour at this meeting.

## Acknowledgements

This work was supported in part by a grant from the Natural Sciences and Engineering Council of Canada and by funding from the Center for Sensor Materials at the University of Notre Dame.

# References

1.  M. Hansen and K. Anderko: *Constitution of Binary Alloys*, McGraw-Hill, New York, 1958, p. 680.
2.  W. A. Deer, R. A. Howie and J. Zussman: *An Introduction to Rock-Forming Minerals*, Longman, London, 1966, p. 346.
3.  E. Tani, M. Yoshimura and S. Somiya, 'Revised Phase Diagram of the System $ZrO_2$-$CeO_2$ Below 1400°C', *J. Am. Ceram. Soc.*, 1983, **66**, 506–510.
4.  T. Takahashi, H. Iwahara and Y. Nagai, 'High Oxide Ion Conduction in Sintered $Bi_2O_3$ Containing SrO, CaO or $La_2O_3$', *J. of Applied Electrochem.*, 1972, **2**, 97–104.
5.  L. G. Sillen and B. Aurivillius, 'Oxide Phases with a Defect Oxygen Lattice', *Z. Krist*, 1939, **101**, 483–495.
6.  B. Aurivillius, 'An X-ray Investigation of the Systems $CaO–Bi_2O_3$, $SrO–Bi_2O_3$ and $BaO–Bi_2O_3$-O (Mixed Oxides with a Defect Oxygen Lattice)', *Arkiv för Kemi, Mineralogi och Geologi*, 1943, **16A** (17), 1–13.
7.  R. Guillermo, P. Conflant, J.-P. Boivin and D. Thomas, 'Le Diagramme des Phases Solides du Systèm $Bi_2O_3$-SrO', *Revue de Chimie Minérale*, 1978, **15**, 153–159.
8.  N. M. Hwang, R. S. Roth and C. J. Brown, 'Phase Equilibria in the Systems SrO–CuO and $SrO-\frac{1}{2}Bi_2O_3$', *J. Am. Ceram. Soc.*, 1990, **73** (8), 2531–2533.
9.  E. A. Payzant and H. W. King, 'An Experimental Evaluation of Computational Methods for Determining Lattice Parameters Using Bragg-Brentano Powder Diffractometry', *Advances in X-ray Analysis*, 1994, **37**, in the press.
10. R. E. Vogel and C. P. Kempter, 'A Mathematical Technique for the Precision Determination of Lattice Parameters', *Acta Cryst.*, 1961, **14**, 1130–1134.
11. H. W. King and E. A. Payzant, 'An Experimental Evaluation of Error Functions for Bragg-Brentano Diffractometry', *Advances in X-ray Analysis*, 1993, **36**, 663–670.
12. C. W. Burnham, 'Lattice Constant Refinement', *Carnegie Institute of Washington Yearbook*, 1962, **61**, 132–135.
13. F. P. Glasser, 'Application of Experimental Techniques and the Critical Determination of Phase Equilibria in Oxide Systems', in *Applications of Phase Diagrams in Metallurgy and Ceramics*, National Bureau of Standards, Gaithersburg, MD, 1977, **STP-496**, 407–422.

# Isotopic Oxygen Exchange and Selection of Materials for Ceramic Ion Transport Membranes

B. C. H. STEELE

*Centre for Technical Ceramics, Imperial College, London SW7 2BP, UK*

## Abstract

A simple model incorporating an interfacial ($R_E$) and bulk ($R_o$) resistive term provides a framework to enable target values ($< 0.15\,\Omega\,cm^2$) for these parameters to be established. The parameter $R_o$ equals the membrane thickness divided by the specific ionic conductivity and so the required thickness for various membranes can easily be specified. The interfacial resistive term incorporates a characteristic surface current density term, $j_E$, which is proportional to the surface exchange coefficient determined by isotopic oxygen exchange experiments. Data are correlated to enable ceramic electrolyte and electrode materials to be recommended for operation over the temperature range 450–950°C.

## Introduction

Many technological devices operating at elevated temperatures incorporate oxide components which are required to sustain high oxygen fluxes. Examples include mixed conductors for ceramic membranes designed to separate oxygen from air, and oxide electrodes for solid oxide fuel cells (SOFC), electrolysers, oxygen pumps and amperometric oxygen monitors. Similarly, oxygen electrolyte materials in these high temperature electrochemical systems are also required to exhibit relatively high oxygen ion conductivities. It should be noted that high oxygen fluxes through these devices also implies large values for the oxygen transport kinetics across the relevant gas–solid interfaces. The magnitude of the oxygen fluxes required for these technological applications has been surveyed elsewhere[1,2] and so emphasis will be given in the present contribution to the role of interfacial reactions.

Oxygen ion fluxes through systems incorporating either mixed conductors or oxide electrolytes may be represented by Fig. 1. The interfacial reactions

$$\tfrac{1}{2}O_2 + V_o^{\cdot\cdot} + 2e' \rightarrow O_o^x \tag{1}$$

can be denoted by a resistive term, $R_E\ \Omega cm^2$.

The magnitude of the electrode resistive term $R_E$ will be determined by the rate controlling step in reaction (1). In principle this rate-limiting process can be due to the Faradaic charge-transfer step, ($R_T$) or to mass transport limitations ($R_C$), or chemical reaction kinetic control ($R_R$). At elevated temperatures and low

49

OXYGEN FLUXES
THROUGH CERAMIC MEMBRANES / ELECTROLYTES

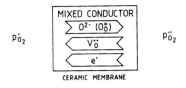

$$\tfrac{1}{2}O_2 + V_O^{\cdot\cdot} + 2e' \longrightarrow O_O^{x} \qquad p'_{O_2} > p''_{O_2} \qquad O_O^{x} \longrightarrow \tfrac{1}{2}O_2 + V_O^{\cdot\cdot} + 2e'$$

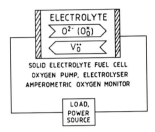

**Fig. 1** *Representation of oxygen ion fluxes through systems incorporating mixed conductors and oxide electrolytes.*

overpotential it is probable that the kinetics of the relevant chemical reaction will be controlling. However, whatever the rate-controlling mechanism very similar expressions can be derived[4] for all the possible processes,

$$R_T = \frac{RT}{zF} \cdot \frac{1}{j_o}, \qquad R_C = \frac{RT}{zF} \cdot \frac{1}{jL}$$

$$R_R = \frac{RT}{zF} \cdot \frac{1}{jR} \cdot \frac{\vartheta}{p}$$

The quantities, $j_o$, $jL$, $jR$ denote the exchange current density, limiting current density and reaction current density respectively. Clearly if the stoichiometric factor ($\vartheta$) and reaction order ($p$) are similar then all the expressions are identical, and it is appropriate therefore to define a 'characteristic' current density, $j_e$, for the specific electrode structure under examination, i.e.

$$R_E = \frac{RT}{zF} \cdot \frac{1}{j_e}$$

The transport kinetics through the impermeable oxide ceramic will also be controlled by another resistive term designated $R_o$ $\Omega cm^2$.

The resistive term, $R_o$ $\Omega cm^{-2}$, will be equal to $L/\sigma$, where $L$ (cm) is the thickness of the oxide ceramic, and $\sigma$ (S cm$^{-1}$), the associated specific ionic conductivity. If $j$ (A cm$^{-2}$) is the oxygen flux being transported through the system then the voltage drop (n, $V$) across the device will be given by,

$$\eta = (R_o + R_E)j \tag{2}$$

i.e. $$j = \eta/(R_o + R_E) \tag{3}$$

or $$j = \eta \, (L/\sigma + RT/j_e zF) \tag{4}$$

In this expression $\eta$ can be replaced by $(RT/zF)\ln pO_2'/pO_2''$, and so equ. (4) is similar to one of the expressions derived by Liu,[5] and if the interfacial resistance $(R_A)$ can be ignored then the expression becomes identical to the known relationship

$$j = [RT/zF)^2]\ln pO_2'/pO_2'' \text{ atom. O cm}^{-2}\text{s}^{-1}$$

derived by Wagner[6] 60 years ago for the parabolic oxidation kinetics of metals.

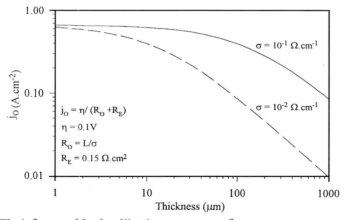

**Fig. 2** *The influence of Surface Kinetics upon oxygen flux.*

The influence of the surface kinetics upon the oxygen flux is depicted in Fig. 2 for a specific value of $R_E$ (0.15 $\Omega$ cm$^2$) and $\eta$ (0.1 V). For a large thickness (1000 $\mu$m) of the membrane the bulk resistance of the membrane dominates but as the thickness of the membrane decreases the surface kinetics become rate-limiting. The membrane thickness at which the change-over from bulk to surface control occurs is given by

$$R_E = R_o$$

i.e.
$$\frac{RT}{zF} \cdot \frac{1}{jE} = \frac{L}{\sigma}$$

$$\therefore \qquad \frac{RT}{zF} \cdot \frac{Vm}{zFk} = L. \frac{Vm\,RT}{D^*(zF)^2}$$

so that $L = D/k$

It has been noted by Kilner[17] that the quantity $D/k$ often has a value around $10^{-2}$ cm (see 'Surface Exchange Mechanisms' p59). This implies that fabricating membranes less than 100 µm thick will not be advantageous unless the value of $k$ can be specifically increased. In the above derivation of $L$, $j_E = zFk/V_m$ (see p. 59), and $\sigma = D^*(zF)^2/V_m RT$ by the Nernst–Einstein equation.

## Selection and Design of Ceramic Membrane Material

### General

Specific ionic conductivity–reciprocal temperature relationships for a selection of oxygen ion conducting electrolytes are depicted in Fig. 3. Assuming a value for $L/\sigma$ equal to 0.15 $\Omega cm^2$, then it is easy to calculate the maximum allowable thickness (right hand ordinate) for a given ionic conductivity value. It will be noted that 150 µm has been taken as the minimum practical thickness likely to be achieved for a self-supported membrane.

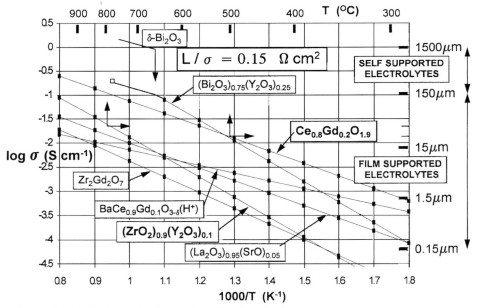

**Fig. 3** *Specific ionic conductivity values for selected oxide electrolytes as a function of reciprocal temperature.*

## Membranes for Operation Above 950°C

The well known cubic zirconia electrolyte, $Zr_{0.85}Y_{0.15}O_{1.925}$ has an ionic conductivity value of $10^{-1}$ S.cm$^{-1}$ at 950°C, and so self-supported ceramic electrolytes with a thickness around 150 µm have been successfully incorporated into planar SOFC modules by many groups around the world, and power densities approaching 0.5 Wcm$^{-2}$ have been achieved. However, operating at 950°C has problems associated with the lack of economic balance-of-plant equipment, and there is a requirement to reduce the operating temperature to the intermediate range 750–800°C (for $CH_4$ fuel) and 450–500°C (for $CH_3OH$ fuel).

## Membranes for Operations at 750–800°C

Further examination of Fig. 3 reveals that the specific conductivity of $Ce_{0.9}Gd_{0.1}O_{1.95}$ electrolytes has a value of $10^{-2}$ Scm$^{-1}$ at 500°C and it is possible, therefore, to envisage using this material as a supported thick film (~ 15 µm) electrolyte structure. Moreover, the electronic conductivity under anodic conditions (~ 0.8V) is much reduced at these lower temperatures and the associated additional losses can be tolerated. The additional development of metallic bi-polar plates by Siemens/Plansee[8] ensures for the first time that individual cells can easily be connected together to assemble a stack operating in the temperature range 450–500°C.

## Interfacial Reactions

### General

It was suggested in the introduction that a typical value for the interfacial resistance, $R_E$, (i.e. $RT/j_e zF$) should be ca. 0.15 ohm.cm$^2$, if power densities approaching 0.5 Wcm$^{-2}$ are to be achieved. At 1000 K this implies a value of $j_e > 3 \times 10^{-1}$ A cm$^{-2}$. This current flux can be written as a molar or volume flux fo $O_2$, i.e.

$$J_e = \tfrac{1}{2} j_e/4F = 4 \times 10^{-7} \text{ mol.cm}^{-2} \text{ s}^{-1} \text{ or } 0.5 \text{ cm}^3 \text{ (STP) } O_2 \text{ cm}^{-2}\text{min}^{-1} \qquad (8)$$

Moreover, the molar flux can be related to the surface exchange coefficient, $k$ (cm.s$^{-1}$), determined by isotopic exchange techniques (12) by the relationship $k = J_o V_m$ where $V_m$ is the molar volume of oxygen at the interface. Whilst the value of $V_m$ is specific to the unit cell dimensions and oxygen stoichiometry of the particular oxide under consideration, a typical value of 20–25 cm$^3$ (for $O_2$) is common for many perovskite oxides. Accordingly, a value of 25 cm$^3$ has been assumed for the present analysis, which results in a value of $k$ approximately equal to $10^{-5}$cm.s$^{-1}$.

Values of the surface exchange coefficient, $k$, can be determined by $^{18}O/^{16}O$ isotopic exchange measurements alone,[9, 10] but we[11] prefer to employ $^{18}O/^{16}O$ isotopic exchange together with a determination of the $^{18}O/^{16}O$ diffusion profile in the solid oxide (IEDP technique). These measurements produce an unambiguous value for the oxygen self-diffusion coefficient, $D_o^*$, and a value for the surface exchange coefficient, $k$, which is specific to the particular surface under examination. Typical data[1] for $La_{0.65}Sr_{0.35}MnO_{3-x}$ and $La_{0.6}Ca_{0.4}Co_{0.8}Fe_{0.2}O_{3-x}$ are reproduced in Figs 4 and 5.

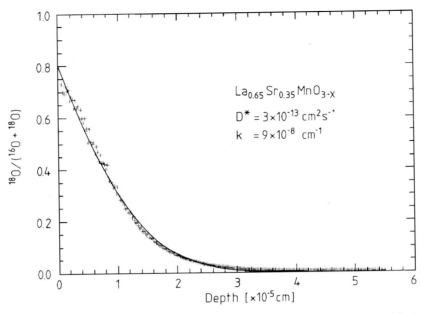

**Fig. 4** $^{18}O/^{16}O$ *isotopic diffusion anneal profile in* $La_{0.65} Mn_{0.35} O_{3-\chi}$ *annealed for 270s at 900°C.*

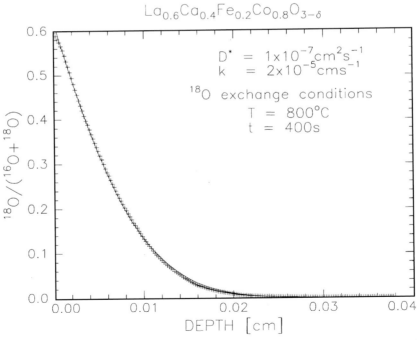

**Fig. 5** $^{18}O/^{16}O$ *isotopic diffusion anneal profile in* $La_{0.6} Ca_{0.4} Fe_{0.2} C_{0.8} O_{3-\delta}$ *annealed for 400s at 800°C.*

It should also be emphasised that the kinetic data values will be a function of the oxide stoichiometry which is controlled by the imposed oxygen partial pressure or overpotential as shown in Fig. 6 for three perovskite compositions. In the temperature range 800–900°C the oxygen vacancy diffusion coefficient, $D_v$ typically has values of $10^{-6}$-$10^{-7}$cm$^2$s$^{-1}$. It follows that the large changes in the oxygen self-diffusion coefficient, $D_o{}^*$, reported in Figs 4 and 5 can be attributed to major differences in the oxygen vacancy concentration, $C_v$ exhibited by these materials (Fig. 6) at high oxygen partial pressures, i.e. $C_o D_o{}^* = C_v D_v$ where $C_o$ is the concentration of oxygen ions.

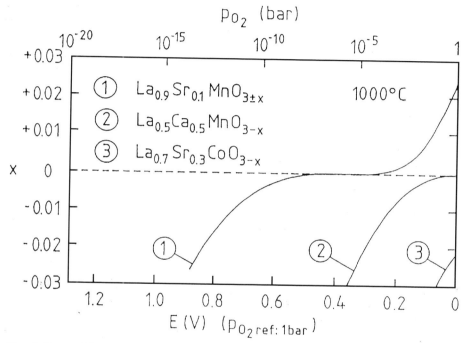

**Fig. 6** *Composition - oxygen partial pressure isotherms for La$_{0.65}$ Sr$_{0.35}$ MnO$_{3\pm x}$, La$_{0.5}$ Sr$_{0.5}$ Mn O$_{3-x}$, and La$_{0.7}$ Sr$_{0.3}$ Co O$_{3-x}$ at 1000°C.*

*Surface Exchange Coefficient Values*

Selected values for the surface exchange coefficient, $k$[9, 10, 11, 12, 13] determined by isotopic exchange experiments are compiled in Fig. 7 together with derived values for the associated exchange current density, $j_o$. These values were all obtained at high oxygen partial pressures and so are applicable to cathodic conditions in SOFC systems.

Examination of Fig. 8 reveals a number of interesting features. The solid electrolyte Zr$_{0.85}$Y$_{0.15}$O$_{1.925}$ has a low surface exchange coefficient, and the deposition of porous Pt electrodes only increases the value of $k$ by about three orders of magnitude. This value is still below exchange coefficient values measured for bare Bi$_{1.6}$Er$_{0.4}$O$_3$ and Ce$_{0.9}$Y$_{0.1}$O$_{1.95}$ electrolyte surfaces. In fact, the magnitude of the

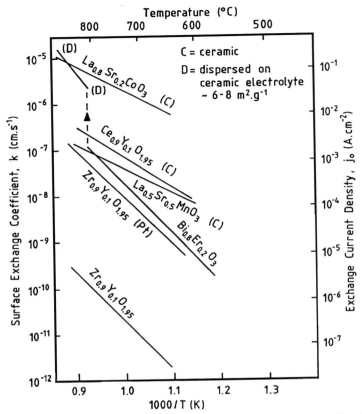

**Fig. 7** *Compilation of oxygen surface exchange coefficient values for selected oxides as a function of reciprocal temperature.*

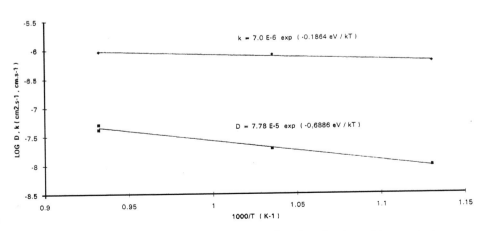

**Fig. 8** D *and* k *Values for* $La_{0.6}Ca_{0.4}Fe_{0.8}Co_{0.2}O_3$ *as a function of reciprocal temperature.*

exchange coefficient for these electrolytes is not influenced by the presence of noble metal (Pt, Au, Ag) electrodes,[12] implying that the electrolyte surface is catalytically active for the dissociative adsorption of oxygen molecules.

To increase the oxygen exchange kinetics of zirconia-based electrolytes it is necessary to use oxide electro-catalysts. Data for bulk $La_{0.5}Sr_{0.5}MnO_3$ samples are included in Fig. 7 and attention is drawn to the fact that the magnitude of the surface exchange coefficient is comparable to values obtained for $Bi_{1.6}Er_{0.4}O_3$ and $Ce_{0.9}Y_{0.1}O_{1.95}$ ceramic electrolytes. However, when fine particles of this oxide are dispersed on zirconia electrolyte surfaces to produce porous electrode structures incorporating a high proportion of three phase boundary regions, then the exchange coefficient increases to around $10^{-6}cms^{-1}$ at 800°C and $10^{-5}cms^{-1}$ at 950°C. However, at 800°C the $k$ value is about an order of magnitude too low and significant cathodic polarisation occurs at this temperature for high current densities.

Ceramic samples of the mixed conductor $La_{0.8}Sr_{0.2}CoO_{3-x}$ exhibit encouraging values for $k$ of ca. $10^{-5}$ cms$^{-1}$ above 800°C, and when these are used in conjunction with zirconia electrolyte, the very high current densities can be sustained with minimal cathodic overpotential.[14] Whilst excellent electro-catalytic performance can be obtained in the laboratory for $La_{0.8}Sr_{0.2}CoO_{3-x}$ for small samples, it is difficult to reproduce this behaviour for large technological samples because the processing routes usually produce high resistance interfacial reaction products, e.g. $La_2Zr_2O_7$, $SrZrO_3$, and the large thermal expansion mismatch between the zirconia electrolyte and the electrode produces cracking and spalling when the electrolyte/electrode structure is cycled.

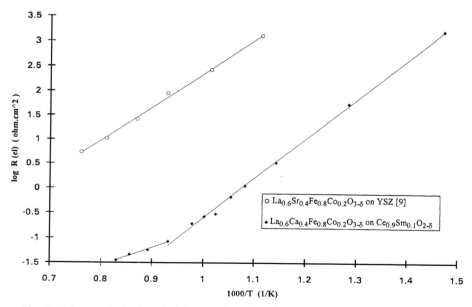

**Fig. 9** *Values of electrode resistivity as a function of reciprocal temperature for $La_{0.6}Sr_{0.4}Fe_{0.8}Co_{0.2}O_{3-x}/Zr_{0.85}{}^N Y_{0.15}O_{1.925}$ and $La_{0.6}Ca_{0.4}Fe_{0.8}C_{0.2}O_{3-x}/Ce_{0.9}Cd_{0.1}O_{1.95}$ cathode / electrolyte assemblies.*

57

It was emphasised earlier (pp 00 and 00) that ceria-based electrolytes such as $Ce_{0.9}Gd_{0.1}O_{1.95}$ could be incorporated in electrochemical systems designed to operate at intermediate temperatures (450–750°C). Ceria-based electrolytes not only exhibit higher oxygen ion conductivities under these conditions but possess further advantages. For example, ceria solid solutions typically exhibit thermal expansion coefficients ca. $12.5 \times 10^{-6}$ $K^{-1}$ (approximately 20% higher than $Zr_{0.85}Y_{0.15}O_{1.925}$). These values are compatible with ferritic stainless metallic components, and also selected La–Co–O electrode compositions, e.g. $La_{0.6}Sr_{0.4}Co_{0.2}Fe_{0.8}O_{3-x}$ (thermal expansion value $14 \times 10^{-6}$ $K^{-1}$). Moreover these electrode compositions appear also to be chemically stable[15] in contact with ceria-based electrolytes as the pyrochlore compound $La_2Ce_2O_7$ does not exist. Although the incorporation of $Fe^{3+}$ lowers the concentration of anion vacancies, samples of $La_{0.6}Sr_{0.4}Co_{0.2}Fe_{0.8}O_{3-x}$ still exhibit relatively high values of $D^*$ and $k$ as shown in Fig. 8. As expected, therefore, electrode resistances associated with $Ce_{0.9}Gd_{0.1}O_{1.95}/La_{0.6}Sr_{0.4}Co_{0.2}Fe_{0.8}O_3$ structures are extremely low as shown in Fig. 9. These values were obtained with symmetrical electrodes[16] using impedance spectroscopy, and it is evident that the electrode resistivities are lower than the target figure of 0.15 $\Omega cm^2$.

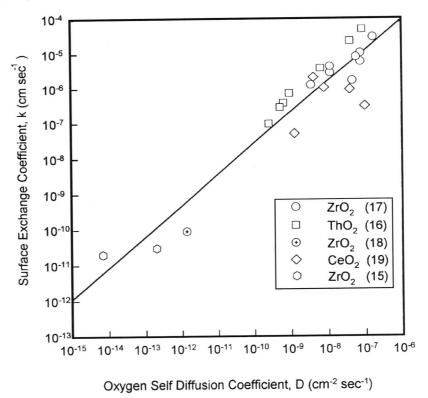

**Fig. 10** *Data for the fluorite oxides* $MO_2$ *showing correlation between K and D values.*

## Surface Exchange Mechanisms

The assumption of a relationship between $k$ and $j_e$ and thus $R_E$ (p00) has been useful in developing alternative cathode materials such as $La_{0.6}Sr_{0.4}Co_{0.2}Fe_{0.8}O_3$. However, these phenomenological relationships provide little information about the mechanisms of surface exchange processes and the associated oxygen electrode reactions.

Attention is drawn to the strong correlation between $k$ and $D^*$ for anion deficient $MO_2$ oxides noted by Kilner[17] and depicted in Fig. 10. The regression line shown in this figure has a slope very close to unity and suggests that the same defect processes, related to the migration and formation of oxygen ion vacancies, are common to both the diffusion and surface exchange process. The value of $k/D$ in Fig. 10 is $1.2 \times 10^2$.

Kleitz et al.[3] have drawn attention to another correlation between $R_E$ and $\rho$ (specific resistivity of electrolyte) and report $R_E/\rho = 3.5 \times 10^{-2}$ for a variety of electrode/electrolyte combinations for $pO_2 = 10^{-4}$ atm. at 780°C. It should be noted that:

$$R_E/\rho = R_E \cdot \sigma = \frac{RT}{zF} \cdot \frac{Vm}{zFk} \cdot \frac{D^*(1/Vm)(zF)^2}{RT} = \frac{D}{k} = 1/1.2 \times 10^2 = 0.83 \times 10^{-2} cm.$$

This value is remarkably close to that mentioned by Kleitz, et al.

It should be emphasised that the isotopic exchange data are obtained for 'bare' electrolyte surfaces whereas the $R_E/1$ values refer to electrolyte surfaces covered with a variety of porous metal electrodes. It could be concluded, therefore, that the electrode resistance $R_E$ is independent of the type of metallic electrode which only functions as a source/sink for electrons. This is indeed the case for $Bi_{1.6}Er_{0.4}O_3$[12] where the catalytic behaviour of the electrolyte surface determines the overall electrode resistance.

However, the situation is more complicated for $ZrO_2$–$Y_2O_3$ electrolytes where the nature of the metallic electrode influences the value measured for the electrode resistance. Silver provides the best electrode kinetic behaviour followed by platinum, then gold. The isotopic exchange data for bare zirconia electrolyte surfaces (Fig. 7) indicates very slow kinetics and noble metals are required for the initial dissociative adsorption step. The slow kinetics for $ZrO_2$ electrolyte surfaces are undoubtedly associated with the relatively small concentration of electronic charge carriers in this material.[16] It is well known[18–20] that the electrode kinetics of $ZrO_2$ electrolytes can be considerably improved by the incorporation of catalytically active cations in the near surface region and presumably this enhanced activity is associated with an increase in the concentration of electronic charge carriers.

Less attention has been given to $CeO_2$-based electrolytes, and available data are confusing probably due to the presence of segregated impurities at interfaces. It should be noted, however, that the isotopic exchange kinetics for $CeO_2$ electrolytes are relatively fast (Fig. 7) probably due to the ability of the $Ce^{4+}/Ce^{3+}$ redox reaction to provide sufficient electronic charge carriers. It is probable, therefore, that the type of electrode material will have less influence on the electrode resistance than in the case for $ZrO_2$ electrolytes.

It is also well known that ceria particles used in three-way automative catalysts function as an oxygen reservoir,[21] and participation of the lattice oxygen in oxide catalysts is well documented in the Max–Van Krevelen mechanism[22] for partial oxidation reactions. A heuristic model[3] appears to be emerging in which the surface oxygen exchange process, and probably the electrolyte/electrode resistivities, are controlled by a reservoir of electroactive oxygen species. This oxygen storage capacity comprises a near surface region and so both diffusive ($D$) and surface exchane ($i$) processes are required to provide the electroactive oxygen species for the Faradaic reaction. To improve the relative electrode kinetics it will be necessary to elucidate how to influence the rate of supply of electroactive species from the reservoir.

## Acknowledgements

The author wishes to thank present colleagues in the Centre for Technical Ceramic for many useful discussions and for providing experimental data. Financial support from the Science and Engineering Research Council, the Commission of the European Communities, and the Club of Industrial Affiliates for the development of Ceramic Electrochemical Reactors is also gratefully acknowledged.

## References

1. B. C. H. Steele: *Mater. Sci. and Eng.*, 1992, **B13**, 79.
2. B. C. H. Steele: *High Temperature Electrochemical Behaviour of Fast Ion and Mixed Conductors*. F. W. Poulson, J. J. Bentzen, T. Jacobsen, E. Skou, and M. J. L. Ostergard. Eds. Riso Nat. Lab., Denmark, 1993, p.423.
3. M. Kleitz, T. Kloidt and L. Dessemond: p.89, ibid.
4. K. J. Vetter: *Electrochemical Kinetics* Academic Press 1967, pp.339-345.
5. M. Liu: p.95. *Ionic and Mixed Conducting Ceramics*, R. A. Ramanarayan and H. L. Tuller, Electrochem. Soc. Inc., New Jersey, 1991, Vol. 91-12.
6. C. Wagner: *Z. Physik. Chem.* 1933, **B21**, 25.
7. M. Dennis-Dumelie, G. Nowogrocki and J. C. Boivin: *Brit. Ceram. Proc.*, 1988, **43**, 151.
8. E. Ivers-Tiffee, W. Wersing, M. Schiessl and H. Greiner: *Ber. Bunsenges. Phys. Chem.*, 1990, **94**, 978.
9. B. A. Boukamp, I. C. Vinke, K. J. De Vries and A. J. Burggraaf: *Solid State Ionics*, 1989, **32/33**, 918.
10. E. Kh Kurumchin and M. V. Perfiliev: *Solid State Ionics*, 1990, **42**, 129.
11. J. A. Kilner, L. Ilkov and B. C. H. Steele: *Solid State Ionics*, 1984, **12**, 89.
12. B. C. H. Steele, J. A. Kilner, P. F. Dennis, A. McHale, M. Van Hemert and A. J. Burggraaf: *Solid State Ionics*, 1986, **18/19**, 1038.
13. S. Carter, A. Selcuk, R. J. Chater, J. Kajda, J. A. Kilner and B. C. H. Steele: *Solid State Ionics*, 1992, **53/56**, 597.
14. Y. Takeda, R. Kanno, M. Noda, Y. Tomida and O. Yamamoto: *J. Electrochem. Soc.*, 1987, **134**, 2656.

15. C. C. Chen, M. M. Nasrallah and H. U. Anderson: in *Proc. of 3rd Intl. Symp. on Solid Oxide Fuel Cells*. Eds. S. C. Singhal and H. Iwahara, eds. Proc. Vol. 93-4, Electrochem. Soc., New Jersey, U.S.A., 1993, p252.
16. B. C. H. Steele: in *Proc. of 3rd Grove Fuel Cell Symposium. J. Power Sources*, 1994, *49*, 1-14
17. J. A. Kilner: *2nd Intl. Symp. Ionic and Mixed Conducting Ceramics*, vol. 94-12, Electrochemical Society, New Jersey, 1994, 174.
18. H. H. Mobius and B. Rohland: U.S. Patent No. 3,377,203 (9.4.68).
19. H. Tannenberger, P. Kovacs. German Patent DE 1771 829 B2 (18.7.68).
20. B. A. Van Hassel and A. J. Burggraaf: *Appl. Phys.*, 1989, **A49**, 33.
21. G. S. Zafiris and R. T. Gorte: *J. Catalysis*, 1993, **139**, 561.
22. B. C. Gates, J. R. Katzer and G. C. A. Schuit: *The Chemistry of Catalytic Processes*. McGraw-Hill, 1979, p.344.

# Gas–Solid Electrolyte Interactions for Chemical Sensing

WERNER WEPPNER

*Sensors and Solid State Ionics, School of Engineering, Christian Albrechts University, D-24098 Kiel, Germany*

## Abstract

The principles of chemical sensing by the interaction of gases with solid ionic conductor surfaces are described. Depending on the exchange process, the sensors are classified as sensors of type I (detection of the ionically mobile component), type II (detection of an immobile component) and type III (equilibration through an auxiliary surface modification). The response is analysed by considering thermodynamic and kinetic aspects. Three types of cross-sensitivities are identified: formation of thermodynamically more stable phases, kinetically more favourable reactions and catalytically active processes. The application of kinetic principles is proposed for more 'intelligent' sensors.

## Introduction

The application of solid ionic conductors for chemical sensors shows a variety of unique features which are advantageous over most other approaches:[1]

1. The ionic conductor converts the chemical energy of reaction into the measured electrical signal. This relates closely to classical chemical analysis in a laboratory by studying chemical reactions.
2. In addition, the reverse process may be applied. Electrical energy may be employed to force a chemical reaction to occur and to study the kinetics of this process.
3. The concentration is directly converted into an easily and very precisely measurable electrical voltage or current signal. No intermediate steps, such as the conversion of an optical into an electrical signal, is required. Voltages and currents are readily displayed or processed for control purposes and automation (robotics).
4. The signal is selective, primarily for the type of mobile species (in the case of solid ionic conductors, only one) and secondly for the chemical species which equilibrate with the surface or junction with the electronic lead.
5. There is hardly any dissolution of other species than the electroactive component in the solid in contrast to liquid electrolytes, which often respond to a large variety of dissolved ions.
6. The electrical potential (emf) is independent of experimental parameters such as size, geometry, crystallinity and even the magnitude of the electrical conductivity.

This type of sensor may be therefore readily miniaturised and integrated into electronic circuits. Solids may be easily machined and remain unchanged in shape.

## Principles

Several principles of chemical sensing by solid ionic conductors may be classified by the type of interaction at the surface.[2]

*Type I sensors:*
The first type of chemical sensor based on solid ionic conductors is a conventional electrochemical concentration cell for the species of interest. The best known example is the λ-probe which makes use of zirconia solid oxygen ion conductors for measuring the oxygen partial pressure or activity.[3] Generally, porous platinum

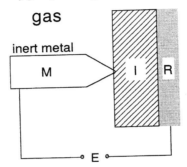

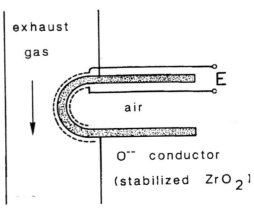

**Fig. 1** *Schematic arrangement of a type I sensor which responds to the mobile species in the solid electrolyte. The best known example is the λ-probe. Electrons are exchanged across the interface between the platinum lead and the solid ionic conductor. The response to oxygen occurs by equilibration of the electrons with the oxygen ions at the surface of the ionic conductor.*

is used as the electronic lead which allows the exchange of electrons with the surface of zirconia which has equilibrated with the oxygen at the given partial pressure (Fig. 1). The measurable voltage $E$ is determined by the Fermi level $E_F$ or electrochemical potential $\eta_e$ of the electrons compared to a reference electrode

$$E = \frac{1}{q} (E_F^{interface} - E_F^{ref}) = \frac{1}{q} (\eta_e^{interface} - \eta_e^{ref}) \tag{1}$$

where $q$ is the elementary charge.

The reference electrode is usually of porous platinum, but in this case with a defined oxygen partial pressure, commonly that of air. Alternatively, mixtures of metals and metal oxides may be employed which define the oxygen partial pressure according to Gibbs's phase rule ($N$ phases of $N$ components at constant total pressure $p$ and temperature $T$). A disadvantage is in this case, however, that any leakage of the electrolyte and the sealing material will result in oxidation or reduction and the oxygen activity may no longer be defined.

The measurement of the pressure or activity of chemical species in type I sensors is the consequence of the equilibration of electrons and ions at the surface of the electrolyte,

$$A^{z_A+} + z_A = A \tag{2}$$

with the equilibrium condition (mass action law)

$$\eta_A z_A + \eta_e^- = {}^\mu A \tag{3}$$

where $\mu$ is the chemical potential. Since the ions are mobile in the electrolyte, there will be no local difference in their electrochemical potential, considering that the flux of ions is given by the gradient of the electrochemical potential,

$$\dot{j}_{A^{z_A}} = - \frac{{}^c A^{z_A}\,{}^D A^{z_A}}{kT} \; \text{grad} \; \eta_{A^{z_A}} = - \frac{{}^c A^{z_A}\,{}^D A^{z_A}}{{}^\mu A \; q^2} \text{grad} \; \eta_{A^{z_A}} \tag{4}$$

where $c$, $D$, $\sigma$, $k$ and $T$ are the concentration, diffusion coefficient, conductivity, Boltzmann's constant and absolute temperature, respectively.

Therefore, the primarily measured difference of the Fermi level is given by the difference in the chemical potential of the neutral mobile component of the electrolyte,

$$E = \frac{1}{z_A q} (\mu_A^{interface} - \mu_A^{ref}) \tag{5}$$

The chemical potential is related to the activity by

$$\mu_A = \mu_A{}^0 + kT \ln a_A \tag{6}$$

and in the case of ideally diluted solutions to the concentrations $c_A$ or partial pressure $p_A$ by

$$\mu_A = \mu_A{}^0 + kt + C_A \ln p_A \tag{7}$$

where $\mu_A{}^0$ is the chemical potential in the standard state ($a_A = 1$).

In this way, type I sensors measure the activity of chemical species indirectly by equilibration with the electrons. Accordingly the surface of the electrolyte should first equilibrate with the gas and then electrons equilibrate with the ions.

*Type II sensors:*

The above derivation of the relation between the cell voltage and the activity indicates that it is not necessary that the mobile ions within the electrolyte have to be identical with the chemical species which are exchanged with the gas, if the mobile and immobile ions of the electrolyte show a well defined relationship. This is the case for a binary compound according to the Gibbs–Duhems relationship

$$\Delta G_f^0 = n_A (\mu_A - \mu_A^0) + n_B (\mu_B - \mu_B^0) \tag{8}$$

where $\Delta G_f^0$, $n_{A,B}$ and $\mu_{A,B}^0$ are the standard Gibbs energy of formation of the electrolyte, the numbers of species $A$ and $B$ and the chemical potentials of the components $A$ and $B$ in the standard state, respectively.

Additional components may be present if these are sufficiently immobile and do not participate in the equilibration; the system may then be considered to be quasi-binary. The following relation holds if the stoichiometry of the solid ionic conductor is nearly the same at both sides in contact with the electrodes:

$$E = \frac{1}{z_B q} (\mu_B^{\text{interface}} - \mu_B^{\text{ref.}}) = \frac{kT}{z_B q} \ln \frac{a_B^{\text{interface}}}{a_B^{\text{ef.}}} \tag{9}$$

If a reference to the mobile species $A$ is used, the relation between the cell voltage and the activity of the chemical species includes the Gibbs energy of formation of the electrolyte:

$$E = - \frac{1}{z_B q} \left[ \mu_B^{\text{interface}} - \frac{\Delta G_f^0}{n_B} + \frac{n_A}{n_B} (\mu_A^{\text{ref.}} - \mu_A^0) + \mu_B^0 \right]$$

$$= - \frac{kT}{z_B q} \ln a_B{}^{\text{gas}} + \frac{\Delta G_f^0}{n_B z_B q} - \frac{n_A kT}{n_B z_B q} \ln a_A{}^{\text{ref.}} \tag{10}$$

66

Sensors based on this principle are called type II sensors in an analogy to the terminology used in conventional electrochemistry of liquids and the expression 'electrodes of 2nd type' which respond by the equilibrium between their components with the electrolyte. Compared to type I sensors the different ions of the electrolyte have to equilibrate with each other in addition to the equilibrium of the ions with the electrons.

The best known example is the measurement of $SO_2/SO_3$ gas pressures by employing mixtures of the cation conductors lithium and silver sulphate.[4] $SO_3$ is exchanged at the surface and determines the chemical potential of silver with reference to elemental silver

$$ E = \frac{kT}{2q} \left[ \ln p_{SO_3}^{\text{interface}} + \frac{1}{2} \ln p_{O_2} - \frac{\Delta G^0_f (Ag_2SO_4)}{kT} \right] \tag{11} $$

It is seen that in this case the electrolyte also equilibrates with oxygen. For $SO_2/SO_3$ measurements the oxygen partial pressure has to be known or needs to be measured separately, e.g. by a type I zirconia based oxygen sensor.

Type I and II sensors represent a quite powerful and successful concept, but lack the availability of suitable solid electrolytes. Type I sensors are presently restricted to oxygen, fluorine, chlorine and possibly hydrogen. Also, solid electrolytes for type I and II sensors require high operating temperatures which are generally much higher than those of the analyte. This may cause shifts in the composition by equilibration and may accordingly lead to erroneous results.

The available low temperature solid electrolytes, such as β-alumina, Nasicon, lithium silicate-phosphate or silver rubidium iodide do not include chemical species of major importance for detection, neither as the mobile nor immobile component. It is possible, however, to make use of these fast solid ionic conductors by employing an auxiliary phase at the surface, which preferentially should be a predominant electronic conductor.

*Type III sensors:[5]*

The principle is illustrated in Fig. 2. The predominantly electronic conductor equilibrates with the chemical species of interest in the surrounding medium and the solid electrolyte measures the chemical potential of one of the other components. In the case of type III sensors the medium equilibrates with a predominant electronic conductor and the leads to the voltmeter are made of an 'ionically conducting wire'. For type III sensors it is not the equilibration between electrons and ions as in type I and II sensors, but the equilibration between two types of ions which controls the voltage. In contrast to type I and II sensors, ionic species are exchanged across the electrolyte/electrode interface in addition to the electrons. Because of the availability of a large number of predominantly electronically conducting compounds there are excellent opportunities to measure many chemical species of practical interest.[6]

The reference electrode at the opposite side of the ionically conducting 'lead' needs to be protected from the surrounding medium which is technologically

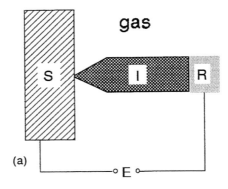

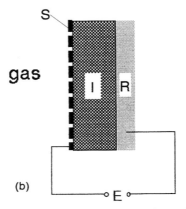

**Fig. 2** *Schematic arrangement of a type III sensor. The surface of the solid ionic conductor is modified by a thin layer of a material which relates the activity of the species of interest to the activity of the mobile component of the solid electrolyte.*

difficult to realise because the surface of the electronic conductor has to be exposed to the medium with the species of interest. Use is therefore made of a thin film of the electronic conductor and the interface with the ionic conductor equilibrates through this thin film. Preferentially, the layer is porous in order to facilitate equilibration. Because of this configuration type III sensors may be also described as surface modified solid electrolyte sensors.

In the case of a ternary system of the two components $M_1$ and $M_2$ and the component of interest $X$, the width of the single phase of the electronic conducting surface modification changes with the activity of $X$, e.g. from the solid to the broken line of the partial Gibbs triangle as shown in Fig. 3. Without any further equilibration with another phase the concentrations of $M_1$ and $M_2$ are at an unknown position along the broken line. In equilibrium with the electrolyte, however, the composition always moves along the border between the single phase and two-phase region. The composition becomes defined in this way and a well defined relation exists between the chemical potentials of all three components. In the general case it is

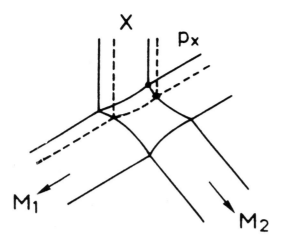

**Fig. 3** *Variation of the phase diagram of a ternary surface modification (solid ➙ broken line) by a decrease of the partial gas pressure. By equilibrium with the solid electrolyte the composition is fixed to the boundary line between the single phase and two-phase region. The activities of both $M_1$ and $M_2$ are fixed by the gas pressure and either one may be measured by the solid electrolyte.*

required that $N$-2 surface modifying phases exist in equilibrium with the solid electrolyte, if the total number of components of all phases is $N$. Together with the surrounding medium all activities are then well defined according to Gibbs's phase rule.

The general relationship between the cell voltage and the activity of the chemical species of interest is derived by considering the Gibbs–Duhem relation for all $N$-1 phases of the surface modification and the solid electrolyte,

$$(\Delta G_{f,p}^0) = \sum_{i=1}^{N} n_{pi}(\mu_i - \mu_i^0) = kT \sum_{i=1}^{N} n_{pi} \ln a_i \qquad p = 1, ..., N\text{-}1 \qquad (12)$$

and making use of Nernst's law for the mobile component (1) in the electrolyte

$$E = -\frac{1}{z_1 q}(\mu_1 - \mu_1^{ref}) = \frac{kT}{z_1 q} \ln \frac{a_1}{a_1^{ref}} \qquad (13)$$

by elimination of all activities except the one to be measured (2):

$$E = \frac{kT}{z_1 qd} \ln a_2 \sum_{p=1}^{N-1} (-1)^{p+1} n_{pi} d_{p1}$$

$$- \frac{1}{z_1 qd} \sum_{p=1}^{N-1} (-1)^{p+1} \Delta G_{f,p}^0 d_{p1} + \frac{kT}{z_1 q} \ln a_1^{ref} \qquad (14)$$

where $n_{pi}$, $d$ and $d_{p1}$ are the number of species of type $i$ in phase $p$, the determinant formed by the stoichiometric numbers of all elements of all $N-1$ compounds, including those of the component which is being measured, and the minor formed by elimination of the $p$-th row and 1st column.

At first glance it appears difficult to prepare the correct $N$ phases which are in equilibrium with each other. In reality, however, nature forms the correct phases itself when the sensor is at the operating temperature and exposed to the medium. It is also rarely necessary to use more than one surface modifying phase even in the case of multiple compounds with more than three components in total. That is because only thermodynamically active components have to be considered which are sufficiently fast and capable of exchange with the medium and the solid electrolyte within the time frame of the measurement. Immobile components are thermodynamically inactive and do not need to be considered. In practice one has to deal in most cases with quasi-ternary systems of the electrolyte $A_{\alpha1}B_{\beta1}C_{\gamma1}$ and the surface modification $A_{\alpha2}B_{\beta2}C_{\gamma2}$:

$$E = \frac{kT\,(\beta_1\,\gamma_2 - \beta_2\,\gamma_1)}{z_1 q\,(\alpha_1\,\gamma_2 - \alpha_2\,\gamma_1)}\, \ln a_2 - \frac{\gamma_2\,\Delta G^0_{f,1} - \gamma_1\,\Delta G^0_{f,2}}{z_1 q\,(\alpha_1\,\gamma_2 - \alpha_2\,\gamma_1)} + \frac{kT}{z_1 q}\, \ln a_1^{\text{ref}} \quad (15)$$

Furthermore, it is not necessary to prepare the surface modification by one of the thin film deposition techniques. This may be readily prepared by exposing the solid electrolyte to the gas and passing a current of ions from the reference electrode to the opposite measuring electrode which is covered by a porous electronic lead and is exposed to the medium. The ions will react with the component which is being measured and form the compound in-situ.

## Experimental Response

Most advanced sensors make use of the available fast ion conductors at ambient temperature, such as $\beta/\beta''$-alumina, Nasicon and silver rubidium iodide. Accordingly, the surface modifications are alkali metal and silver salts with anions corresponding to the analysed component. In some cases, additional components are included to improve the performance such as the stability against reaction with water. Commonly, these components are thermodynamically inactive. Surface modified type III sensors have so far been studied for the measurement of mono-elemental gases $Cl_2$ and $O_2$ (at lower temperatures than is typical for type I oxygen sensors)[7] and a variety of multi-elemental gases such as $SO_2$, $CO_2$ and $NO_x$.[8]

The response to $CO_2$ will be taken as an example to show the general features. In this case, the sensor interacts with two gases, $CO_2$ and $O_2$, as shown schematically in Fig. 4. Thermodynamically, the cell voltage depends on both partial pressures:

$$E = E_o + \frac{kT}{2q}\, \ln p_{CO_2} + \frac{kT}{4q}\, \ln p_{O_2} \quad (16)$$

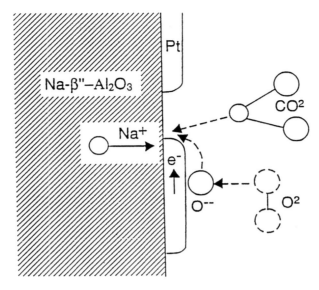

**Fig. 4** *Illustration of the galvanic cell reaction of a $CO_2$ sensor. The electroactive sodium ions have to react simultaneously with $CO_2$, $O_2$ and the electrons from the electronic lead in order to provide the voltage corresponding the Gibbs energy of formation of $Na_2CO_3$ at the given $CO_2$ and $O_2$ gas pressures. Any partial reactions which include the electrons will result in lower cell voltages. Any further reaction at a later stage will be a common chemical reaction and has no effect on the cell voltage.*

The formation of the carbonate is a 2-electron process for the reaction with the $CO_2$ molecules and a 4-electron process for the reaction with the $O_2$ molecule. The voltage change per decade of the partial pressure change of $O_2$ is therefore only half that of $CO_2$.

The cell voltage as given by eq. (16) is only observed if the reaction of the electroactive species occurs with both $CO_2$ and $O_2$ in one single step. Consecutive steps will partially result in the transformation of chemical energy into heat rather than electrical energy. This will cause a lower cell voltage. Only the part of the reaction which includes the transfer of electrons from the electronic lead to the ions of the electrolyte counts for the cell voltage. Assuming that the electroactive ion and the oxygen form an oxide in the moment when the electron is provided by the electronic lead, the cell voltage will correspond to the Gibbs energy of formation of the oxide from the metal with the activity of the reference electrode and oxygen of the given partial pressure. Any further reaction to form the cabonate from the oxide and the $CO_2$ in the gas phase does not involve electrons from the electronic lead and is therefore not 'seen' by the cell voltage. As in a common reaction, the chemical energy of this part of the reaction is converted into heat. In reality it is often observed that several reactions occur in parallel. These processes contribute in various amounts to the cell voltage and an intermediate value is experimentally observed.

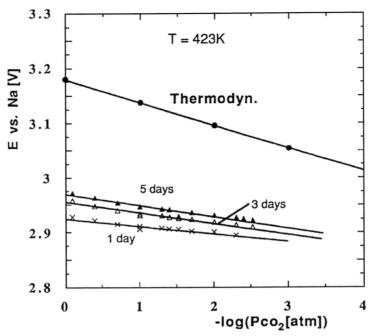

**Fig. 5** *Theoretical and experimental response of a CO₂ sensor at 150 °C and an oxygen partial pressure of 0.2 atm. A drift of the voltage toward the theoretical values is observed in the course of time.*

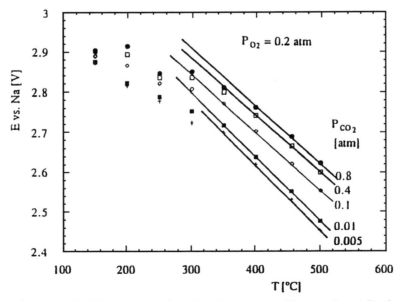

**Fig. 6** *Response of a CO₂ sensor as a function of temperature. The experimentally observed cell voltages approach the theoretical ones at a temperature above about 300 °C.*

For the detection of $CO_2$ the best results so far were observed by the application of Na–$\beta$–$Al_2O_3$ as a fast sodium ion conductor and $Na_2CO_3$ as surface modification. The observed cell voltage as a function of the $CO_2$ pressure is shown in Fig. 5 for a temperature of 150°C and a fixed oxygen partial pressure of 0.2 atm. For comparison the thermodynamically expected voltage is shown. This clearly has a higher absolute value and a higher slope. In addition, the experimental values show a slow drift of several mV per day toward the thermodynamically more stable values. The difference may be explained by the formation of 23% $Na_2O_2$ and 77% $Na_2CO_3$ at the beginning.

Fig. 6 shows the results obtained at higher temperatures in the range 300–550°C. Above 400°C the formation of $Na_2CO_3$ in a single step is no longer impeded. This behaviour is independent of geometric parameters and preparation conditions. Reproducible values are obtained without calibration. Only the knowledge of the oxygen pressure is required.

The interaction with oxygen of the gas phase depends on various experimental parameters. Any response from no sensitivity to oxygen at all up to a 4-electron process has been observed. Nasicon generally shows a smaller response to oxygen than Na-$\beta$/$\beta''$–$Al_2O_3$.

## Cross Sensitivity

Cross sensitivity to other species of the medium under investigation is generally the most restrictive problem of any present sensor development. Several quite different interfacial phenomena occur which range from the formation of thermodynamically more stable phases, e.g. $Na_2SO_4$, to the formation of kinetically more favourable compounds. At partial pressures sufficiently below those of $CO_2$, cross sensitivities are commonly negligible because of their small contribution to the actual galvanic cell reaction. The cross sensitivities may be classified into three types:[9]

1. *Thermodynamic formation of more stable phases.* The medium includes in this case species which form thermodynamically more stable compounds with the electroactive ions than the species which are of interest. This reaction generally increases the cell voltage with reference to an electrode of the elemental electroactive component. As an example, Fig. 7 shows the interaction of a $Cl_2$ sensor with $SO_2$ based on a silver ion conductor. For various temperatures the line is given which separates the more favourable conditions of formation of $Ag_2SO_4$ (above) and AgCl (below). This diagram indicates that even extraordinarily small traces of $SO_2$ would be sufficient to form $Ag_2SO_4$ at low temperatures. In reality, however, the electroactive species will react with those species which are actually available in sufficiently large numbers at the electrode. As long as $SO_2$ is in the minority there will be little influence in spite of the high stability of the sulphate. It is also possible to increase the temperature in order to avoid the influence of $SO_2$ since the chloride becomes increasingly more stable with increasing temperature.

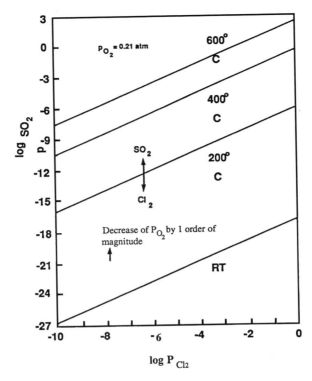

**Fig. 7** *Theoretical interaction of the presence of $SO_2$ in the case of a $Cl_2$ sensor with AgCl as surface modification. At the partial pressures above the lines which are given for various temperatures the formation of $Ag_2SO_4$ is thermodynamically more favourable, whereas AgCl is more stable below.*

2. *Kinetic formation of more stable phases.* The interaction with some species of the medium may be kinetically preferred though the reaction should not occur because of thermodynamic reasons. An example is the formation of $Ag_2S$ in the case of the presence of $H_2S$ in spite of the much lower stability of $Ag_2S$ than AgCl. This interaction results in a decrease of the cell voltage as shown in Fig. 8. For practical applications this interaction is highly undesirable because it gives credence that the chlorine partial pressure is much lower than is actually the case. The formation of $Ag_2S$ is favoured because of the extraordinary fast chemical diffusion coefficient of this compound[10] and therefore the possibility of very fast growth.

3. *Catalysed surface reactions.* The higher temperature of the sensor than that of the medium and the presence of catalytically active materials, e.g. noble metals, may result in reactions of various species of the medium which are thermodynamically favourable but kinetically impeded. As an example, chlorine shows a charge transfer reaction with water in the presence of platinum as electrode material as shown in Fig. 9. This reaction decreases the chlorine activity

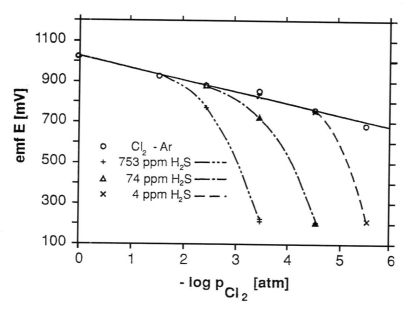

**Fig. 8** *Influence of H₂S on a chlorine sensor based on a solid silver ion conductor in spite of the much lower thermodynamic stability of Ag₂S as compared to AgCl. At approximately the same concentration of Cl₂ and H₂S the effect becomes remarkable when silver ions arriving at the interphase with the electrode 'do not see' the Cl₂ and form the kinetically more favourable Ag₂S which provides a lower cell voltage. Any later formation of the more stable AgCl is a chemical reaction and has no influence on the voltage.*

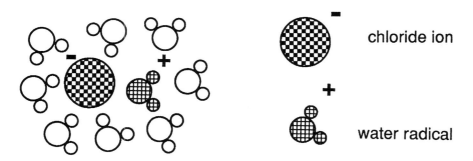

**Fig. 9** *Charge transfer reaction of chlorine and water at the heated platinised surface of the sensor.*

and accordingly also decreases the cell voltage. The influence is quite small, however, as shown in fig. 10 because of the small energy of formation of the complex. In other cases, e.g. in the presence of reducing species when oxygen partial pressures are measured, the energy of interaction may be much larger and has a strong effect on the cell voltage.

75

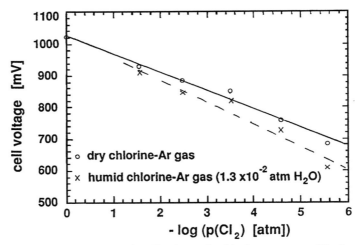

**Fig. 10** *Effect of water vapour on the cell voltage of a chlorine gas sensor. The formation of the thermodynamically favourable charge transfer complex shows a decrease of the $Cl_2$ activity which corresponds to a small decrease in the cell voltage.*

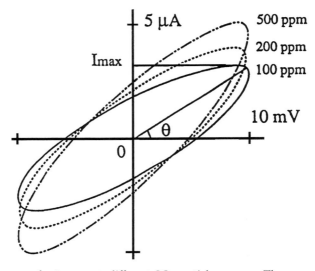

**Fig. 11** *Response of a $\theta$-sensor to different $CO_2$ partial pressures. The current-voltage curve shows a hysteresis and phase shift due to the kinetics of the galvanic cell reaction of formation of $Na_2CO_3$. Various parameters, such as the maximum current or the angle of the main axis of the ellipsoid, may be taken as an indication of the $CO_2$ pressure.*

## Kinetic Principles

Potentiometric sensors as discussed above may be regarded as 'passive' sensors since it is simply *observed* which reactions occur between the electroactive ion and the medium. Solid electrolytes may also play an 'active' role in the galvanic cell reaction. A current may be passed through the electrolyte at various rates and the ions are *forced* to react at this rate with the available constituents of the medium.[11] For each component different electrode processes occur which may be further specified by applying different reaction rates. In an extraordinarily simple arrangement the solid electrolyte is covered on both sides with the surface modification or with an inert, porous metallic lead and a periodically variable voltage is applied. Fig. 11 shows the current voltage curve for various $CO_2$ partial pressures. For the analysis of $pCO_2$ a variety of parameters may be considered, e.g. the angle of the main axis of the ellipsoidal figure or the maximum current. This type of sensor is called a $\theta$-sensor. There is no separate reference electrode required, but a calibration of each individual sensor is necessary. The shape of the signal may even be interpreted in further detail for the concentration of various other components.

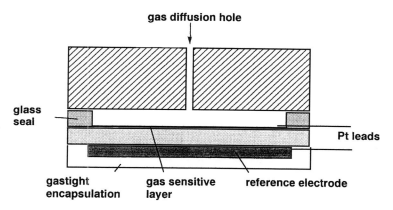

Fig. 12 *Limiting current chlorine sensor based on the transport of silver ions across a solid ion conductor and the reaction with chlorine which has diffused through a thin hole in a cover into a cavity at this side of the electrolyte.*

Furthermore, the amperometric principle of diffusion limited barriers may be employed.[12] As an example, in the case of chlorine, a sensor with a solid silver ion conductor is employed which transports silver from an encapsulated reference electrode into a cavity which is in contact with the gas by a small hole in the cover (Fig. 12). The silver reacts with the chlorine which is consumed in this way as the result of diffusion of chlorine through the small hole. Without any current, the potentiometric voltage is observed. With a decrease of this voltage, the current is increased until the limiting case of negligible chlorine concentration in the cavity and maximum concentration gradient across the hole is reached (Fig. 13). This limiting current is accordingly proportional to the chlorine partial pressure. In contrast to the logarithmic relationship of a potentiometric sensor, a linear relationship is obtained which allows much higher precision for smaller variations of the partial pressure.

## Conclusion

The combination of potentiometric and amperometric measurements is an extraordinarily powerful tool which is unique to solid electrolytes. The combination of these parameters allows limitations experienced by other types of sensors to be overcome.

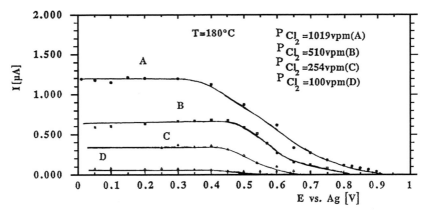

**Fig. 13** *Response of a limiting current chlorine sensor. If no current is passed, the voltage agrees with that of a potentiometric sensor. With decreasing voltage the electrical current carries silver ions into the cavity and consumes the chlorine gas until the concentration gradient of chlorine through the hole takes up a maximum and diffusion is limiting the current. The plateau current is proportional to the chlorine partial gas pressure.*

## References

1. W. Weppner: 'Advanced Principles of Sensors Based on Solid State Ionics', *Mat. Sci. and Eng. B*, 1992, **15**, 48–55.
2. W. Weppner: 'Solid State Electrochemical Gas Sensors', *Proc. 2nd Int. Meet. Chem. Sensors*, J.-L. Aucouturier, J.-S. Cauhapé, M. Destrian, P. Hagenmuller, C. Lucat, F. Ménil, J. Portier and J. Salardienne eds, Bordeaux, 1986; *Sensors and Actuators*, 1987, **12**, 449-453.
3. H. Dietz, W. Haecker and H. Jahnke: *Adv. Electrochem. Electrochem. Eng.*, H. Gerischer and C. W. Tobias eds, J. Wiley & Sons, New York, NY, 1977, Vol. 10, p1.
4. W. L. Worrell and Q. G. Liu: *Anal. Chem. Symp. Ser.*, 1983, **17**, 332; *J. Electroanal. Chem. Interfacial Electrochem.*, 1984, **168**, 355.
5. W. Weppner: *German Patent*, No. 2926 172 (28 June 1979); *US Patent* 4.352.068 (28 Sept. 1982); 'Surface Modification of Solid Electrolytes for Gas Sensors', *Solid State Ionics*, 1990, **40/41**, 369–374.
6. W. Weppner: 'Thin Film Solid State Ionic Gas Sensors', *Solid State Microbatteries*, J. R. Akridge and M. Balkanski eds, Plenum Press, New York, NY, 1990, pp395–405; G. Hötzel, Ph.D. Thesis, Stuttgart University, 1988.

7.  A. Menne and W. Weppner: 'Selectivity, Response Time and Chemical Aspects of Gas Sensing AgCl Layers for Gas Sensors', *Solid State Ionics*, 1990, **40/41**, 468–471; 'Redox Reaction Mechanisms at the AgCl/Chlorine Gas Interface', *Electrochim. Acta*, 1991, **36**, 1823–1827.

8.  G. Hötzel and W. Weppner: 'Potentiometric Gas Sensors Based on Fast Solid Electrolytes', *Sensors and Actuators*, 1987, **12**, 107–119; J. Liu and W. Weppner: 'β"-Alumina Solid Electrolytes for Solid State Electrochemical $CO_2$ Gas Sensors', *Solid State Communications*, 1990, **76**, 311–313; 'Potentiometric $CO_2$ Gas Sensor Based on Na-β/β"-Alumina Solid Electrolytes at 450°C', *Eur. J. Solid State Inorg. Chem.*, 1991, **28**, 1151–1160.

9.  W. Weppner: 'Development of Thin Film Surface Modified Solid State Electrochemical Gas Sensors', *Recent Advances in Fast Ion Conducting Materials and Devices*, B. V. R. Chowdari, Q.-G. Liu and L.-Q. Chen eds, World Scientific, Singapore, 1990.

10. W. F. Chu, H. Rickert and W. Weppner: 'Electrochemical Investigations of Chemical Diffusion in Wustite and Silver-Sulphide', *Fast Ion Transport in Solids; Solid State Batteries and Devices*, W. van Gool ed., North-Holland, Amsterdam–London, American Elsevier, New York, NY, 1973, pp181–191.

11. J. Liu and W. Weppner: 'θ-Sensors: A New Concept for Advanced Solid State Ionic Gas Sensors', *Applied Physics A*, 1992, **55**, 48–55.

12. B. Y. Liaw, J. Liu, A. Menne and W. Weppner: 'Kinetic Principles for New Types of Solid State Ionic Gas Sensors', *Solid State Ionics*, 1992, **56**, 18–23; J. Liu and W. Weppner: 'Limiting Current Chlorine Gas Sensor Based on β"-Alumina Solid Electrolyte', *Sensors and Actuators B*, 1992, **6**, 270–273.

# Miniaturised Sulphur Oxide Integrated Sensor

C. D. Eastman

*Alberta Microelectronic Centre, University of Alberta, Edmonton, Canada T6G 2T9*

T. H. Etsell

*Department of Mining, Metallurgical and Petroleum Engineering, University of Alberta, Edmonton, Canada T6G 2G6*

M. J. Brett

*Department of Electrical Engineering, University of Alberta, Edmonton, Canada T6G 2G7*

## Abstract

Work is being performed on the fabrication of an integrated miniature $SO_2$ sensor. Advantages of miniaturisation are described. A solid electrolyte was in the form of sintered discs of $Li_2SO_4$–$Ag_2SO_4$. Initial research was carried out with a sintered disc of $Ag/Li_2SO_4$–$Ag_2SO_4$ as the reference electrode and a Pt gauze working electrode. This cell has been tested over a wide range of air–$SO_2$ mixtures. At 535°C:

$$\text{E.M.F. (mV)} = 216.1 + 73.2 \log (ppm\ SO_2) \quad R^2 = 0.995$$

from 11.6 to 1737 ppm $SO_2$. Subsequently, a miniaturised $SO_2$ sensor was constructed with a reference electrode consisting of a thick film of $Ag/Li_2SO_4$–$Ag_2SO_4$ brushed onto the solid electrolyte substrate. Utilising vacuum evaporation, thin films of platinum (100 nm) were deposited onto the electrolyte discs forming working electrodes with a high concentration of triple point sites. Stable voltages have been recorded for over 4 months for air–$SO_2$ mixtures with 400 to 1000 ppm $SO_2$.

## Introduction

Sulphur dioxide is emitted in large quantities by the combustion of fossil fuels (coal and oil) and is probably the most important compound in the definition of the acidic properties of the atmosphere. This study reports on the development of a miniaturised $SO_2/SO_3$ chemical sensor which will utilise thick and thin film technology. The miniaturised sensor is primarily intended for the measurement of stack gas atmospheres, and for improved control of chemical and metallurgical recovery processes which contain sulphur dioxide/trioxide. This sensor would also be suitable for monitoring air quality.

Through the experience gained in constructing a sulphur dioxide/trioxide sensor, additional sensors for nitrous dioxide and carbon dioxide detection, for instance, could be fabricated. Other solid electrolytes displaying oxygen, silver, sodium, potassium, hydrogen or copper conduction are also suitable candidates for miniaturisation and integration.

## Solid Electrolyte Materials

The solid electrolyte utilised in this study was a solid oxysalt consisting of 75 mol % lithium sulphate ($Li_2SO_4$) and 25 mol % silver sulphate ($Ag_2SO_4$). As investigated previously by Worrell[1] and Mari *et al*,[2] this solid electrolyte has to date proven to be superior to others for $SO_2/SO_3$ detection. In general, solid electrolytes have been found to be significantly less sensitive to the effects of other chemical species and therefore tend to be more selective with respect to chemical potentials measured. However, investigations by Gauthier and Chamberland[3] have shown that sulphur containing gas species such as hydrogen sulphide and mercaptan substances were indistinguishable from an equivalent amount of $SO_2$ with this electrolyte. It is suggested that their removal could be affected with a Ag wool scrubber which would allow differentiation among the sulphur-containing compounds.[4] Another possibility for separation of $SO_2$ and $SO_3$ from other gases would be to use dense mixed conducting sulphate membranes.

## Miniaturisation

Reasons for miniaturisation of the sensor are as follows:

1.  By decreasing the thickness of the electrolyte the response time of the sensor can be reduced as was reported by Mari *et al*.[2] The minimum electrolyte thickness used in their investigation was in the order of 2000 μm. By utilising a planar structure this dimension can ultimately be reduced to 2–10μm, or in the case of thin ceramic films, less than 1 μm.
2.  The use of porous sputtered platinum thin film electrodes greatly enhances the kinetics of the electrode reactions. Previous investigations by both Mari *et al*.[2] and Imanaka *et al*.[5] have proven the advantages of utilising sputtered thin films of platinum. At the electrodes a series of reactions occur at so called 'triple points' (consisting of the gaseous phase, platinum and solid electrolyte) and, by sputtering a porous thin film, the number of these sites is dramatically increased resulting in improved electrode kinetics. The most serious limitations of solid state electrolysers and electrochemical sensors are excessive voltage losses and a lack of sensitivity, respectively, at low temperatures. Utilising an electrolyte thick/thin ceramic film along with a porous platinum thin film, which would form the working electrode, the operating temperature could be significantly lower.
3.  A micron sized thick (platinum or ruthenium) or thin film (nickel-chromium) heating element attached under the supporting substrate would result in power requirements in the order of a few hundred milliwatts. An integrated sensor would then be possible with the addition of a thermocouple or R.T.D. for

temperature measurement and control of the heater. In addition, a smart sensor could be built through the addition of the appropriate algorithms for control of temperature and gas flow rates, measurements of voltage and diagnostic routines for determination of cell functionality.

## Gas–Solid Sensor Theory

The sensor consists of a solid state reference electrode in combination with the working electrode which is exposed to the gas flow to be analysed. For this sensor the following reactions apply:

$$SO_3 + \frac{1}{2}O_2 + 2e^- \leftrightarrow SO_4^{2-} \qquad \text{Working electrode (cathode)}$$

$$\frac{2Ag + SO_4^{2-} \leftrightarrow Ag_2SO_4 +}{2Ag + SO_3 + \frac{1}{2}O_2 \leftrightarrow Ag_2SO_4} \qquad 2e^- \text{ Reference electrode (anode)}$$

$$2Ag + SO_3 + \frac{1}{2}O_2 \leftrightarrow Ag_2SO_4 \qquad \text{Overall sensor reaction} \qquad (1)$$

Utilising the Nernst equation given by:

$$E = E° - \frac{RT}{nF} \ln Q \qquad (2)$$

where $E$ is the measured voltage, $E°$ is the voltage of the overall reaction at standard conditions, $R$ the gas constant, $T$ the temperature in K, $n$ the number of electrons transferred, $F$ the Faraday constant, and $Q$ is the reaction quotient for the overall reaction, the voltage of the sensor should be equivalent to:

$$E = E° + \frac{RT}{2F} \ln \frac{(P_{O_2})^{1/2} (P_{SO_3})}{a_{Ag_2SO_4}} \qquad (3)$$

In this case $E°$ is the voltage of overall reaction (1) under standard conditions at the operating temperature and $n = 2$. The partial pressure of oxygen $P_{O_2}$ is equivalent to 0.2095 atm due to the air in the air–$SO_2$ mixture (for the mixtures used, the $SO_2$ content is low enough so that this assumption is true). The activity of $Ag_2SO_4$ has been determined by previous researchers to be equal to its mole fraction which is 0.25 for this sensor.

For the partial pressure of sulphur trioxide ($P_{SO_3}$) it can be shown the relationship to the partial pressure of sulphur dioxide ($P_{SO_2}$) is given by:

$$P_{SO_3} = \cfrac{P_{SO_{2(in)}}}{\left( 1 + \cfrac{1}{K_p (P_{O_2})^{1/2}} \right)} \tag{4}$$

$P_{SO_{2(in)}}$ is the partial pressure of the sulphur dioxide in the incoming gas stream, $K_p$ is the equilibrium constant for $SO_2 + \frac{1}{2}O_2 \leftrightarrow SO_3$ at the operating temperature (535°C), $P_{O_2}$ is the partial pressure of oxygen in the gas stream, and $P_{SO_3}$ is measured at the working electrode.

## Construction and Assembly of the Sensor

The sensors were constructed with a reference electrode consisting of Ag powder mixed with the electrolyte (25 mol % $Ag_2SO_4$ and 75 mol % $Li_2SO_4$). Initially, in a 2 to 1 weight ratio of silver to $Ag_2SO_4$-$Li_2SO_4$, this layer was 2 mm thick by 25.4 mm in diameter. Onto this layer 2 mm of solid electrolyte powder was pressed (see Fig. 1). The two layers were then pressed in a carbide-lined steel die at 280 MPa for 2–3 h. The pellet was then sintered in an atmosphere of 500 ppm $SO_2$ in air for 2 days at 535 °C. For this sensor, platinum gauze forms the working electrode.

Subsequently, the reference electrode powder mixture was in turn mixed with an organic carrier. The powder content and chemical composition was adjusted so as to obtain the optimum viscosity and flow characteristics for the method of application. At this point the thick film was brushed onto solid electrolyte substrate discs that had been pressed and sintered as described. It would also be possible to utilise screen printing or spinner technology for deposition of the reference electrode materials.

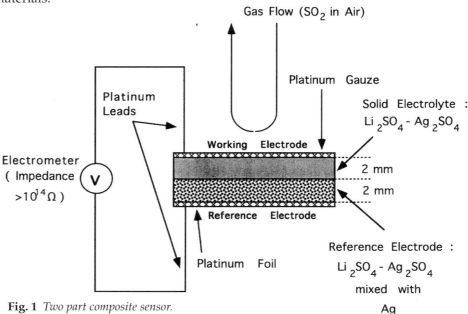

**Fig. 1** *Two part composite sensor.*

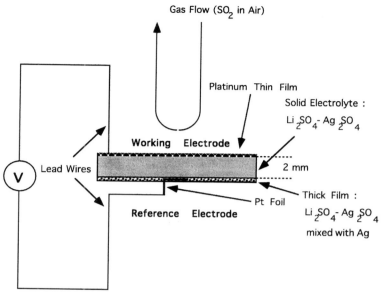

**Fig. 2** *Thick film reference electrode sensor.*

From previous research with pure silver paints deposited onto $Ag_2SO_4$–$Li_2SO_4$ substrates, lifetimes of 2 weeks were observed after which the sensor voltage would rapidly decline. Accordingly, a reference electrode consisting of a 2 : 1 weight ratio of silver to $Ag_2SO_4$–$Li_2SO_4$ was deposited onto the $Ag_2SO_4$–$Li_2SO_4$ solid electrolyte substrate (as shown in Fig. 2). On the other side of the substrate a thin film of platinum was now deposited (vacuum evaporation) forming the working electrode. A platinum ring was placed against the working electrode for electrical contact. The lifetimes were observed to be in the order of 4–5 weeks. In response to these observations the weight ratio at the reference electrode was reduced to 1 : 1 of silver to $Ag_2SO_4$–$Li_2SO_4$.

The sensors were mounted in a machined ceramic fitting fabricated from Macor (Corning) and placed inside an alumina tube exiting a furnace at both ends allowing fittings for gas exchange, thermocouples and electrical leads.

The mixture of $SO_2$ and air was passed through a quartz tube with a catalytic chamber containing cut up platinum gauze and foil. This chamber assures that the $SO_2$ is converted to $SO_3$. At the working electrode, platinum gauze contacts the electrolyte or a platinum ring contacts the thin film platinum layer as previously mentioned, whereas solid platinum foil contacts and seals the reference electrode.

For automated control of the furnace and sensor, a computer control and data acquisition system was assembled. This system consists of a Macintosh IIci running a combination of hardware and software (LabView) from National Instruments. Custom software with a window environment and graphical instrument control was written for control and voltage measurement of the gas sensor. A set of electronic mass flow controllers (Advanced Semiconductor Material-model AFC 260) controlled by the computer system was utilised for mixing of the $SO_2$ in air. For measurement of the sensor temperature, a set of K-type thermocouples (Omega Engineering) was placed on either side of the sensor. The voltage leads from the sensor and thermocouples were connected to the computer via an analog to digital board. All measurements of voltage were taken with 16 bit precision through this board.

## Experimental Results

In Fig. 3 the electromotive force (mV) vs. the concentration of $SO_2$ (ppm) in air is plotted for the composite sensor. The best fit line from these values was determined to be:

$$\text{E.M.F. (mV)} = 216.1 + 73.2 \log (\text{ppm } SO_2) \quad R^2 \text{ (correlation coefficient)} = 0.995 \quad (5)$$

This equation is valid for the range from 11.6 to 1737 ppm $SO_2$ in air.

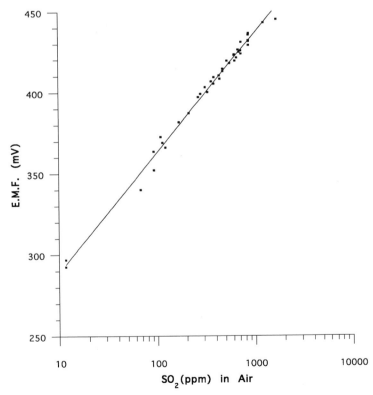

**Fig. 3** *Electromotive force vs. concentration of $SO_2$ in air.*

In Fig. 4 a check for the effect of flow rate with the 900 ppm mix is displayed. From about 40 sccm to 125 sccm the voltage falls within ± 2 mV. This is more than adequate for a sensor of this type.

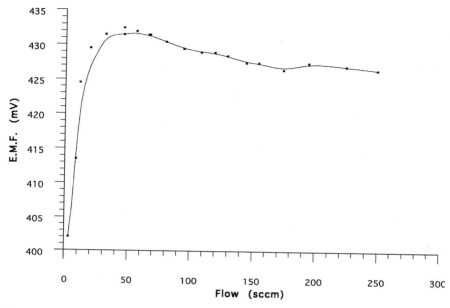

**Fig. 4** *Electromotive force vs. flow rate for 900 ppm SO₂ in air.*

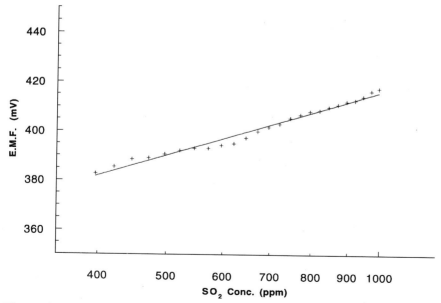

**Fig. 5** *Electromotive force vs. concentration of SO₂ in air.*

87

The data presented in Fig. 5 are from a thick film reference electrode sensor which had been running for over 4 months. It has been observed that there has been no change in the sensor voltage at various $SO_2$ concentrations since startup. The best fit line from these values was determined to be:

$$E.M.F. \ (mV) = 158.1 + 86.0 \log (ppm \ SO_2) \quad R^2 \ (correlation \ coefficient) = 0.991 \quad (6)$$

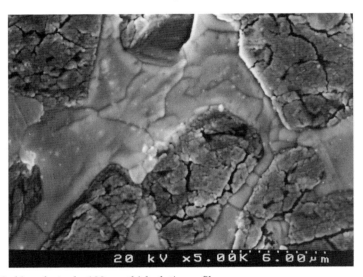

**Fig. 6** *Working electrode, 100 nm thick platinum film.*

In Fig. 6 a photograph of the working electrode consisting of a platinum thin film 100 nm thick deposited on the solid electrolyte substrate is shown. The platinum is clearly resting on top of the solid electrolyte in the form of globular islands. In addition there is considerable microcracking within the platinum islands which would also be beneficial as additional triple point sites. This structure possesses the optimum geometry for catalysis of the conversion of $SO_3$ to $SO_4^{2-}$ in the electrolyte. It maximises the number of triple point sites along the platinum–electrolyte–gas phase interface. Typically, in the deposition of thin films, the parameters required for stress-free films fall within a narrow region of pressure, deposition rate, substrate temperature and surface roughness. Since increasing the density of triple points is advantageous, strict control of deposition parameters is not required. In effect fabrication of the platinum thin film with the required properties is simplified.

## Discussion

The gas–solid composite sensor displayed excellent performance in the range from 11.6 to 1737 ppm of $SO_2$ in air. It has given repeatable response with values obtained

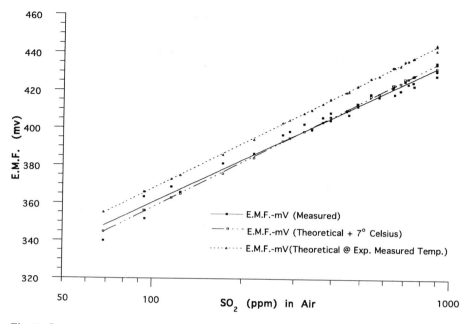

**Fig. 7** *Comparison of measured and calculated emf values with and without temperature adjustment.*

within ~2.5% of theoretical, and has proven to be faster than the thicker gas–gas sensor design.

In the construction of the sensor a thermocouple is placed about 5–10 mm in front of the electrode in the gas stream. For this reason there may be a temperature differential between the thermocouple and the electrode surface. Taking the measured temperatures, a set of emf values was calculated utilising the theoretical equations for the gas concentrations used in the experiments. Another set is calculated in the same manner except 7°C is added to the measured temperatures. Comparison of the calculated and experimentally determined values is shown in Fig. 7. If the theoretical values are assumed to be accurate, then the shift of the measured values from theory (calculated to be only ~ 2.5% for the entire range) is probably due primarily to errors in temperature measurement. In the future either further thermcouples or thin/thick film resistive temperature devices (R.T.D.) will be added for precise measurement of temperature.

The thick film reference electrode gas–solid sensor has displayed excellent performance in the range from 400 to 1000 ppm of SO₂ in air. It has displayed a response with values which are consistently within ±0.5 % from a baseline. However, agreement with theoretical voltages was poorer than for the composit sensor. It is assumed at this point that this can be attributed to the differences in the way in which the reference electrode was constructed with the thick film as compared to a two-part pellet construction. There is also undoubtedly some error in measurement of sensor temperature. These factors warrant further investigation.

The thin film of platinum was adherent to the solid electrolyte throughout the test; however, since the sensor is loaded into a large cylindrical ceramic chamber, examination of response times to changes in gas concentration was not possible. Any response times observed were a function of the purging of the chamber in contact with the working electrode. Nevertheless, it was observed that the voltage would begin changing immediately once the gas concentration was altered. The remainder of the response time was required to purge the chamber to the desired gas concentration. In the future a small gas chamber will be utilised along with platinum thin film thicknesses between 50 and 500 nm so as to determine response times to changes in gas concentration.

## Acknowledgement

Funding was provided by the Alberta Oil Sands Technology and Research Authority and the Alberta Microelectronic Centre.

## References

1. W. L. Worrell: 'Solid–Electrolyte Sensors for $SO_2$ and/or $SO_3$', in *Chemical Sensor Technology*, T. Seiyama ed., Elsevier, Amsterdam, 1988, Vol. 1. p. 97.
2. C. M. Mari, M. Beghi, S. Pizzini and J. Faltemier: 'Electrochemical Solid-state Sensor for $SO_2$ Determination in Air', *Sensors and Actuators B*, 1990, **2(1)**, 51–55.
3. M. Gauthier and A. Chamberland: 'Solid-State Detectors for the Potentiometric Determination of Gaseous Oxides-I. Measurement in Air', *J. Electrochem. Soc.*, 1977, **124**(10), 1579–83.
4. D. Zhang, Y. Maeda and M. Munemori: 'Chemiluminescence Method for Direct Determination of Sulfur Dioxide in Ambient Air', *Anal. Chem.*, 1985, **57**(13), 2552–55.
5. N. Imanaka, G. Adachi and J. Shiowaka: 'Sulfur Dioxide Gas Detection with $Na_2SO_4$-$Y_2(SO_4)_3$-$SiO_2$ Solid Electrolyte by a Solid Reference Electrode Method', *J. Electrochem. Soc.*, 1985, **132(10)**, 2519–20.

# Developments in Solid Electrolyte Sensors for On-Line Measurements

DEREK J. FRAY

*Department of Mining and Mineral Engineering, University of Leeds, Leeds LS2 9JT, UK*

## Introduction

In spite of the fact that solid electrolytes have been known for more than 100 years, it is only recently that the possible technological impact of these materials has been realised. As early as 1908,[1] solid ionic conductors were being used to determine thermodynamic data. In the 1930s, Tubandt and his co-workers[2] confirmed that many halides were ionic conductors but the applications were restricted to simple cells operating over a limited temperature range due either to the low melting points or the occurrence of non-stiochiometry at high temperatures. In many of these salts, both the cations as well as the anions were mobile, requiring that the electrodes were reversible to two different ionic species. Following this active period, research effort diminished until 1957 when Kiukkola and Wagner,[3,4] Ure[5] and Liddiard[6] re-stimulated interest in high temperature galvanic cells by using stabilised zirconia in which only the oxygen ion is mobile. This and related oxygen electrolytes have found industrial application as sensors for the determination of oxygen in molten steel, copper, the air-fuel ratio in internal combustion engines and furnaces and, in the laboratory, as electrolytes in fuel cells and electrocatalytic reactors.

In the past 30 years, environmental issues and the needs for the conservation and storage of energy have lead to the discovery of a large number of other solid electrolytes which may eventually find application in batteries and fuel cells. In particular, the discovery of the high ionic conductivity of sodium beta-alumina[7] and silver rubidium iodide,[8] both of which have conductivities approaching that of fused salts, has led to the synthesis of other 'superionic' conductors.

## Sensors Based on Stabilised Zirconia

This paper is concerned with the application of solid electrolytes to the detection of elements in molten metals at elevated temperatures. The basis of the measurement is derived directly from the Nernst equation which for the cell:

$$Mo : Cr, Cr_2O_3 / CaO \cdot ZrO_2 / \underline{O} : Mo \qquad (1)$$

is

$$- ZEF = RT \ln P_{O_2} (ref) / P_{O_2} \qquad (2)$$

where CaO·ZrO$_2$ is stabilised zirconia in which the conduction process is by the migration of oxygen ions, $Z$ is the charge transferred, $E$ is the measured potential, $F$ is Faraday's constant, $R$ is the gas constant, $T$ is the temperature, $P_{O_2}$(ref) is the equilibrium partial pressure over a mixture of Cr/Cr$_2$O$_3$ at temperature $T$ and $P_{O_2}$ is the partial pressure of oxygen in equilibrium with the oxygen dissolved in the molten metal. If the reference pressure and the equilibrium constant between the partial pressure of oxygen and the activity in the melt are known, the activity can be calculated and from the thermodynamics of the solution the concentration can be derived. It, therefore, appears to be relatively easy to select an electrolyte which conducts an ion of the species to be measured and material which will give a fixed partial pressure or activity of the species. The overall arrangement can be very simple, as shown in Fig. 1 and the method offers the possibility of giving

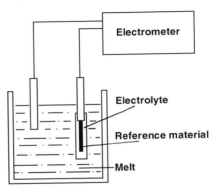

**Fig. 1**  *Schematic of sensor arrangement.*

instantaneous readings which can be interfaced with microprocessors for the control of processes. There are other additional benefits which can accrue from knowing the analysis at the end of the refining reaction. The operator does not need to wait for the result so that further processing can take place and the melt has minimum interaction with the environment and the refractories.

As is mentioned above, oxygen measurements are commonplace in the copper and steel industries[9–11] but the number of other chemical sensors in commercial use in extractive metallurgy is negligible and, furthermore, under certain conditions the performance of the zirconia-based electrolytes can be unsatisfactory. This can be seen in Fig. 2, where reliable data are obtained at high oxygen concentrations but at the lower concentrations, which might be found in steel and copper, after deoxidation, the data are very scattered. The reason for this is that at low oxygen concentrations, the stabilised zirconia no longer acts as a pure ionic conductor but now has an electronic contribution to the overall conductivity and the simple Nernst Eq. (1) no longer applies. This is due to oxygen in the stabilised zirconia dissolving in the melt, leaving vacant oxygen sites and electrons according to the equation

$$O_O = V_O^{\cdot\cdot} + 2e' + \underline{O} \tag{3}$$

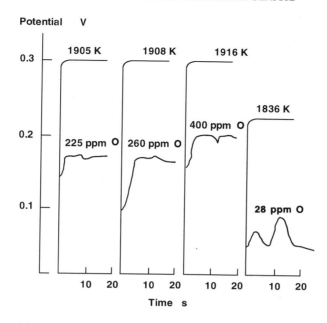

**Fig. 2** *Potential versus time for oxygen sensors in steel.*[10]

where $O_O$ is oxygen on a oxygen lattice site in the stabilised zirconia, $V_O^{\cdot\cdot}$ is a vacant oxygen site in the lattice, $e'$ are electrons and $\underline{O}$ is the oxygen dissolved in the molten metal. If the fraction of the ionic conductivity to the total conductivity is known, the Nernst equation can be altered to take account of this effect by incorporating the transport number for the oxygen ion into the equation. In practice, it is not quite as simple, because the presence of electronic as well as ionic conductivity means that oxygen in the melt can combine with the electrons and diffuse through the material to the reference. This obviously results in mass transfer of the oxygen and can result in the interface concentration of oxygen varying with the mass transfer conditions in the melt. As these are not accurately known in a given melt it is impossible to precisely predict the influence on the cell potential. There is, therefore, a need for sensors which will respond to oxygen at low concentrations in copper and steel that has been deoxidised.

There are many other systems, such as lead and tin, in which it might be useful to be able to measure the oxygen content. Belford and Alcock[12,13] determined the relevant thermodynamic data using stabilised zirconia as an electrolyte, so that all the scientific information is available. However, although it would be of industrial interest to measure oxygen in tin and lead, there is no record of this being attempted.

## Sensors based on other Solid Electrolytes

There are a myriad of solid electrolytes which have been discovered and investigated, mainly from the viewpoint of energy storage. From the on-line sensor point of view, one of the more interesting groups of materials is that of the beta aluminas.[7] Sodium beta alumina, which has a very high ionic conductivity due to the mobility of the sodium ions, has much promise as an electrolyte in high temperature storage batteries and it also offers the possibility for measuring sodium in metallic systems in which sodium is an important addition or impurity. During the pyrometallurgical extraction of zinc, arsenic and antimony are co-reduced and report in the zinc. These elements are generally removed by the addition of sodium to form sodium arsenic and antimony compounds. The sodium is usually added periodically and there is little control over this process as, at the present time, it is not possible to measure the sodium content instantaneously. A sodium sensor has been developed using sodium beta alumina as the electrolyte.[14] The reference material to use is not entirely obvious as, unlike the oxygen sensors where a simple mixture of metal and a metal oxide can give fixed oxygen partial pressure, most sodium-containing intermetallics are either unstable or undergo oxidation at elevated temperatures. One possible answer is to use a mixture of sodium-containing oxide compounds, which are both ionic and electronically conducting, and then to fix the oxygen partial pressure.[15] From the phase rule, a mixture of sodium ferrite and ferric oxide or sodium beta alumina and alpha alumina would give a constant activity of sodium oxide at a given temperature. If the oxygen pressure is held constant either by exposure to air or a metal–metal oxide mixture, this will fix the activity of the sodium in the reference. Industrial trials have shown that this is a feasible way of controlling the sodium content of zinc. Similar trials, using the same design of sensor, are being undertaken to control the modification of aluminium–silicon alloys and to monitor the removal of sodium from aluminium–magnesium alloys.

One of the properties of the beta alumina type structure is that it is very accommodating for a large variety of cations. In some cases, these can be prepared directly, while in other examples, the sodium can be exchanged for another cation in a fused salt and this gives the basis for a range of sensors, silver, copper, lithium, which might find application in parts of the non-ferrous industry.[7]

*Measurement of Oxygen in Molten Metals*

The discussion of the reference material for sodium sensors leads directly to a method of measuring of oxygen in molten metals at very low concentrations. Sodium beta alumina has negligible electronic conductivity over a very wide range of conditions. If the electrolyte is made two phase; for example, a mixture of alpha and beta alumina, the activity of sodium oxide is fixed. When sodium beta alumina is immersed in a molten phase, one might expect that it would respond to sodium. However, in some molten phases, sodium has never been detected; for example, molten steel and copper at elevated temperatures. The electrolyte then attempts to come to equilibrium with some other species in the melt. If oxygen is present, it seems logical that the following equilibrium will be established:

$$Na_2O_{\text{(beta alumina)}} = \underline{O}_{\text{(melt)}} + 2Na \tag{4}$$

The equilibrium constant is given by:

$$K = a_O \cdot a_{Na}^2 / a_{Na_2O} \tag{5}$$

It can easily be seen that any change in the activity of oxygen results in a change in the activity of sodium which can be detected by the sodium beta alumina electrolyte. An extension of this reasoning leads to the conclusion that there is no need for the sodium ferrite reference so that a two-phase electrolyte is required together with a constant oxygen pressure, which could be air. Sensors, based on these concepts, have been used to determine low oxygen contents in molten steel, copper and aluminium.[16]

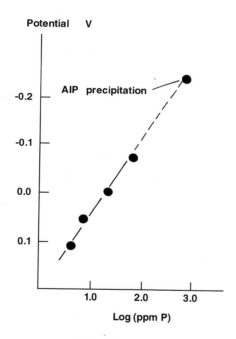

**Fig. 3** *Potential versus parts per million of phosphorus in aluminium.*

*Measurement of other Species*

The above concept can be extended to the measurement of other species in molten metals using a variety of electrolytes. For example, by incorporating a layer of zirconia and zirconium silicate, which fixes the activity of silica, it has been shown that it is possible to measure silicon in molten steel using stabilised zirconia.[17] Effectively what happens is that the silica dissociates into silicon and oxygen:

$$SiO_2 = \underline{Si} + 2\underline{O} \tag{6}$$

The equilibrium constant, $K$, is given by:

$$K = a_{\underline{Si}} \cdot a^2_{\underline{O}} / a_{SiO_2} \tag{7}$$

and any variation in the silicon activity will result in a change in the activity of oxygen which is detected by the electrolyte. This approach can be used for many systems by incorporating the oxide of the element of interest with stabilised zirconia.

A similar approach has been adopted by Alcock who used a composite strontium fluoride–lanthanum fluoride electrolyte with sulphide or oxide additions to measure sulphur and oxygen.[18,19] Beta alumina can be used,[20] for example, if there is a need to measure antimony in zinc, in which sodium antimonate can be mixed with the sodium beta alumina to form a composite electrolyte. In contact with the zinc the following equilibrium will be set up:

$$NaSbO_3 = Na + \underline{Sb} + 3\underline{O} \tag{8}$$

and the equilibrium constant is:

$$K = a_{Na} \cdot a_{\underline{Sb}} \cdot a_{\underline{O}}^3 / a_{NaSbO_2} \tag{9}$$

As the activity of the sodium antimonate is unity, the activity of sodium will be influenced by both the antimony and oxygen contents of the melt. At first sight, it may appear that it is necessary to measure the oxygen content in the zinc, but it is found that this is constant and is given by the zinc/zinc oxide equilibrium. If the melt contains sodium, the sensor would obviously respond to the sodium content of the melt rather than the antimony, although in many cases the two elements could be in equilibrium. If this is not the case, the antimony can easily be measured by using potassium antimonate and a potassium conducting electrolyte. It can, therefore, be seen that by taking account of the other elements in the melt and judicious selection of the electrolyte and the supporting compound, it is possible to measure virtually any element that is present. In dilute solutions it is important to consider the stability of the compound in relation to the other elements that are present. For example, if sulphur is present in combination with oxygen and sulphur needs to be measured, it would be more logical to select a beta alumina or other electrolyte, in which the sulphide of the cation is more stable than the oxide. Sensors, based on these concepts, are undergoing industrial trials for the determination of sodium, arsenic and antimony in zinc, sulphur, phosphorus and oxygen in copper and sodium in aluminium.[21] Results for the measurement of phosphorus in molten aluminium are shown in Figure 3.

*Measurement of High Concentration of Species*
Sensors based upon the Nernst Eq. (1) are remarkably sensitive at low concentrations due to the logarithmic form of the expression. At high concentrations, where the species might be present in the percent of tens of percent range, the scatter in the voltage reading is likely to exceed the predicted change in potential. One such

example where there is a need to measure the change in concentration at high concentrations is copper in matte smelting. This paper has concentrated, to a large extent, on electrolytes which come from the beta alumina family but, in the case of copper, it is impossible to make tubes of copper beta alumina directly as the phase is unstable well below the sintering temperature. In order to overcome this problem, copper zirconium phosphate was synthesised and found to be an excellent copper ion conductor over a wide range of temperatures.[22] In order to measure copper, at high concentrations, it was decided that instead of measuring the potential as in the case of a sensor based upon the Nernst equation, to apply a potential and measure the current. On the application of the current, copper is ionised at the surface of the electrolyte, passes through the electrolyte, and deposits on the cathode. If the volume of matte or metal directly in contact with the electrolyte is kept stagnant, the only way the copper, transported through the electrolyte under the imposed potential, can be replaced is by diffusion through the stagnant layer of matte or metal. If the concentration at the interface is virtually zero the current will depend upon the flux of copper diffusing through the stagnant matte layer to the interface and this will be directly proportional to the concentration in the bulk matte. So instead of measuring a potential which is logarithmically dependent on the concentration, the current is measured and this is linearly dependent on the concentration. This sensor is shown in Fig. 4 and the results in Fig. 5.

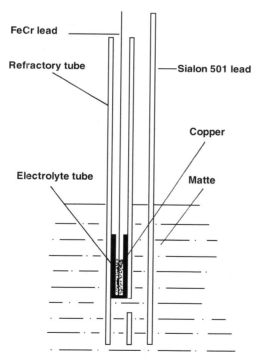

**Fig. 4**  *Sensor for measuring the concentration of copper in matte.*

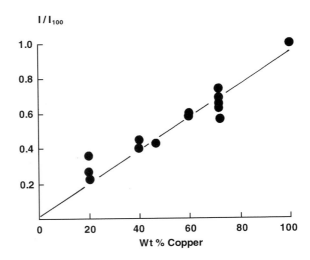

**Fig. 5**  *Fraction of current versus weight percent copper in matte at 1473K.*

## Discussion

This paper has demonstrated that it is possible to measure most elements that occur in mattes, metals and slags using solid electrolytes. Whereas the first solid electrolytes that were applied industrially used an electrolyte which conducted the ion of the species of interest and used a reference to give a fixed activity of the species on the inside of the sensor, it is now possible, by the selection of a second phase, to extend the application of this technique to a wide variety of elements. In certain instances, the concentration of the species is very high and sensors, based upon the Nernst equation, are not suitable. A coulometric sensor can be used in which the current is measured instead of the potential. Obviously, this can only be applied to elements for which a solid electrolyte exists but the method could be used for mattes and alloys containing an appreciable quantity of the element.

This approach opens up the possibility of being able to measure virtually any element that is present in solution in molten metals provided that there is careful selection of the electrolyte and the second phase. It is surprising that this type of sensor has not found wide industrial application. To a certain extent this may be due to the fact that many of the industrialists with the problem to solve are either unaware of the problem or alternatively of its solution. Another difficulty is that in order for an operator to become confident with a new measuring device, a considerable number of plant trials are necessary. These can never be organised at short notice and care has to be taken to avoid disruption of the production processes coupled with alienation of the workforce. It therefore takes considerable effort to build up confidence in both the device and also the people demonstrating it. Fortunately for those developing sensors but, perhaps, less fortunate for the possible users is that there are considerable pressures for better quality, better control, and less pollution and one way of achieving these goals is to have continuous monitoring of the processes. Solid electrolyte sensors, built upon the fundamental work of the 1950s and 1960s,[3,4,7,12,13] offer the possibility of meeting this challenge.

## References

1.  M. Katayama: 'Cells with Solid Electrolytes of the Type of Amalgam Concentration Cells, Chemical Cells and Daniell Cells,' *Z. Physik. Chem.*, 1908, **61**, 566–571.
2.  C. Tubandt, H. Reinhold and W. Jost: 'Ion Migration in Mixed Crystals and its Relationship to Migration in Pure Salts,' *Z. Physik. Chem.*, 1927, **29**, 69-78.
3.  K. Kiukkola and C. Wagner: 'Galvanic Cells for the Determination of the Standard Molar Free Energy of Formation of Metal Halides, Oxides and Sulphides at Elevated Temperatures,' *J. Electrochem. Soc.*, 1957, **104**, 308–21.
4.  K. Kiukkola and C. Wagner: 'Measurements on Galvanic Cells involving Solid Electrolytes,' *J. Electrochem. Soc.*, 1957, **104**, 379–387.
5.  R. W. Ure: 'Ionic Conductivity of Calcium Fluoride Crystals,' *J. Chem. Phys.*, 1957, **26**, 1363–1373.
6.  A. B. Lidiard: 'Anion Diffusion in Alkali Halide Crystals,' *Phys. Chem. Solids*, 1958, **6**, 298–300.
7.  Y. F. Y. Yao and J. T. Kummer: 'Ion Exchange Properties and Rates of ionic Diffusion in β-Alumina,' *J. Inorg. Nuc. Chem.*, 1967, **29**, 2453–2475.
8.  J. N. Bradley and P. O. Greene: 'Potassium Iodide and Silver Iodide Phase Diagram. High ionic Conductivity of $RbAg_4I_5$,' *Trans. Farad. Soc.*, 1966, **62**, 2069–2075.
9.  H. Tominaga, K. Yajima and S. Nishiyama: 'Application of Oxygen Probes for Controlling Oxygen Content in Molten Copper,' *Solid State Ionics*, 1981, **3/4**, 517–574.
10. E. T. Turkdogan and R. J. Fruehan: 'Review of Oxygen Sensors for Use in Steelmaking and of Deoxidation Equilibrium,' *Can. Met. Quart.*, 1972, **11**, 371–384.
11. C. K. Russell, R. J. Fruehan and R. S. Rittiger: 'Probing for More than Temperature,' *J. Met.*, 1971, **23** (11), 44–47.
12. C. B. Alcock and T. N. Belford: 'Thermodynamics and Solubility of Oxygen in Liquid Metals from EMF Measurements involving Solid Electrolytes Part 1 Lead,' *Trans. Farad. Soc.*, 1964, **60**, 822–833.
13. T. N. Belford and C. B. Alcock: 'Thermodynamics and Solubility of Oxygen in Liquid Metals from EMF Measurements Involving Solid Electrolytes Part 2 Tin,' *Trans. Farad. Soc.*, 1965, **61**, 443–453.
14. D. J. Fray: 'Determination of Sodium in Molten Aluminum and Aluminium Alloys using a Beta Alumina Probe,' *Met. Trans.*, 1977, **8B**, 153–156.
15. R. J. Brisley and D. J. Fray: 'Determination of the Sodium Activity in Aluminum and Aluminum Silicon Alloys using Sodium Beta Alumina,' *Met. Trans.*, 1983, **14B**, 435–440.
16. R. V. Kumar and D. J. Fray: 'Application of Novel Sensors in the Measurement of Very Low Oxygen Potentials,' In Press. *Solid State Ionics*.

17. M. Iwase: 'Rapid Determination of Silicon Activities in Hot Metal by Means of Solid State Electrochemical Sensors Equipped with an Auxiliary Electrode,' *Scand. J. Metals*, 1988, **17**, 50–56.
18. C. B. Alcock and B. Li: 'A Fluoride-Based Composite Electrolyte,' *Solid State Ionics*, 1990, **39**, 245–249.
19. C. B. Alcock and L. Wang: 'A Lanthanum Fluoride Based Composite Electrolyte for use as an Oxygen Sensor,' *High Temperatures-High Pressures*, 1990, **22**, 449–451.
20. D. J. Fray: 'Developments in On-line Sensing in Molten Metals using Solid Electrolytes,' *Benelux Metallurgie*, 1990, **39** (3), 63–68.
21. L. J. Cobb, A. Davidson, L. C. Zhang and D. J. Fray, Unpublished work.
22. P. C. Yao and D. J. Fray: 'The Preparation and Properties of the Solid State Ionic Conductor, $CuZr_2(PO_4)_3$,' *Solid State Ionics*, 1983, **8**, 35–42.

# Nonstoichiometry and the NiAs Structure

H. Ipser, K. L. Komarek and R. Krachler

*Institute of Inorganic Chemistry, University of Vienna, Austria*

### Abstract

A comprehensive model for the NiAs-structure is formulated which allows the thermodynamic activity of a component over the entire range of homogeneity to be described, irrespective of the defect mechanism. The equations are applied to activities in the Ni–Te, Fe–Te, Fe–Se, Fe–S, Pd–Te and Co–Sb systems.

## Statistical Model

Several authors have in the past developed theoretical equations to describe the thermodynamic properties in non-stoichiometric NiAs-phases as a function of composition.[1] However, no equation was applicable over the entire range of non-stoichiometry. In the following a single and unique equation is derived which can describe the activity of one component over the entire range of homogeneity, independently of the type of NiAs-phase, i.e. partially filled up or defected, as long as the 'defects' (vacancies or interstitial atoms) are statistically distributed over the respective lattice sites. Thus this equation should be applicable equally well for NiAs-phases with transition metal deficiency (as in the chalcogen systems) or with transition metal surplus (as in transition metal systems with elements of group 13, 14, and also 15); and they should also cover, of course, the 1 : 1 stoichiometry.

In principle the hexagonal NiAs-lattice can be divided into the transition metal sublattice (α-sublattice) and the sublattice formed by the main group element (β-sublattice). In the ideal, perfectly ordered stoichiometric compound $AB$ ($A$ = transition element, $B$ = main group element) all α-sites will be filled by A-atoms and all β-sites by B-atoms. Additional A-atoms enter the trigonal-bipyramidal (or interstitial) positions (thus approaching the $Ni_2In$-structure type). These interstitial positions form an additional sublattice, the i-sublattice. If on the other hand A-atoms are removed from the α-sublattice, then we have two possibilities: vacancies can be formed on all α-sublattice sites, or they can be restricted to alternate layers, thus approaching the $CdI_2$-structure type (where alternate transition metal layers are completely empty). To allow for this latter situation, the α-sublattice is divided into two parts (consisting of alternate transition metal layers), the α1- and the α2-sublattice.

For the derivation of the model we assume that the β-sublattice remains undisturbed; this is certainly justified, because the deviation from the 1 : 1 stoichiometry is always caused by addition or removal of A-atoms. Since a perfectly ordered lattice can only exist at absolute zero, there must always be some disorder at finite temperatures, even at the stoichiometric composition, i.e. a certain number of A-atoms will leave the α-sublattice and enter interstitial positions. To take care of the possibility that atoms can be removed preferentially from alternate layers we consider the *two* equilibria:

$$A_{\alpha 1} \rightleftharpoons A_i$$
$$A_{\alpha 2} \rightleftharpoons A_i$$

Following Wagner and Schottky,[2] we define the concentrations of defects at stoichiometry as the so-called disorder parameters, *a* and *b*, which are constant for any specific NiAs-phase at a given temperature:

$$a = (N_\square^{\alpha 1}/N^t)_{stoich}$$
$$b = (N_\square^{\alpha 2}/N^t)_{stoich}$$

with  $N_\square^{\alpha 1}$ = number of vacancies on the α1-sublattice

$N_\square^{\alpha 2}$ = number of vacancies on the α2-sublattice

$N^t$ = total number of lattice sites, excluding the interstitial sublattice,

i.e. $N^t = N^\alpha + N^\beta = N^{\alpha 1} + N^{\alpha 2} + N^\beta$.

*a* and *b* are a measure for the tendency to form vacancies on the α1- or α2-sublattice, respectively. Of course, the concentration of interstitial A-atoms at stoichiometry, $(N_A^i/N^t)$, must be equal to *a* + *b*.

If *a* = *b*, then the tendency to form vacancies will be the same for both sub-sublattices α1 and α2, and the vacancies will be distributed statistically over all transition metal lattice sites. If one of the parameters becomes very small or even practically zero, then we will have vacancies only in alternate layers, and with decreasing transition metal content the lattice will approach the $CdI_2$-type structure.

The total Gibbs energy of a crystal with NiAs-structure can now be constructed for any composition by adding up the different contributions

$$G = (1/N_L) \, (N^t\mu^* + N_A^i\mu_A^i + N_\square^{\alpha 1}\mu_\square^{\alpha 1} + N_\square^{\alpha 2}\mu_\square^{\alpha 2}) - TS_{conf}$$

$\mu^*$ = Gibbs energy of an ideally ordered crystal of one mole ($A + B$) with stoichiometric composition,

$\mu_A^i$ = change of Gibbs energy, referred to one mole, for the insertion of A-atoms into interstitial sites.

$\mu_\square^{\alpha 1}, \mu_\square^{\alpha 2}$ = change of Gibbs energy for the removal of 1 mole of A-atoms from the α1- or α2-sublattice, respectively. All these different Gibbs energy terms are assumed to be independent of composition.

Keeping in mind that the β-sublattice remains undisturbed, the configurational entropy is (according to Boltzmann's formula) given by

$$S_{conf} = k \ln \quad \frac{N^{\alpha 1}!}{N_\square^{\alpha 1}! \, (N^{\alpha 1} - N_\square^{\alpha 1})!} \quad \frac{N^{\alpha 2}!}{N_\square^{\alpha 2}! \, (N^{\alpha 2} - N_\square^{\alpha 2})!} \quad \frac{N^i!}{N_A^i! \, (N^i - N_A^i)!}$$

When thermodynamic equilibrium is established, the following equilibrium conditions must be valid (at constant temperature and pressure), as shown by Wagner and Schottky:

$$A_{\alpha 1} \rightleftharpoons A_i$$

$$\left( \frac{\partial G}{\partial N_A^i} \right) dN_A^i + \left( \frac{\partial G}{\partial N^{\alpha 1}} \right) dN_\square^{\alpha 1} = 0; \quad dN_A^i = dN_\square^{\alpha 1}$$

$$A_{\alpha 1} \rightleftharpoons A_i$$

$$\left( \frac{\partial G}{\partial N_A^i} \right) dN_A^i + \left( \frac{\partial G}{\partial N^{\alpha 2}} \right) dN_\square^{\alpha 2} = 0; \quad dN_A^i = dN_\square^{\alpha 2}$$

With these equations and with the relationship between the mole fraction and the numbers of the different defects

$$N_A = (1-x_B)N_L = (N^t/2) - N_\square^{\alpha 1} - N_\square^{\alpha 2} + N_A^i$$

we have the basic tools to calculate (by some involved algebra) their composition dependence for the given parameters $a$ and $b$. With the well known thermodynamic relationships

$$\mu_A = RT \ln a_A = N_L \left( \frac{\partial G}{\partial N_A} \right)_{N_B,\, p,\, T}$$

$$\mu_B = RT \ln a_B = N_L \left( \frac{\partial G}{\partial N_B} \right)_{N_A,\, p,\, T}$$

we can obtain the thermodynamic activity of A and B in the following form:

$$\ln a_A = \ln a_{A,o} + \ln \frac{(x_B - 2n_\square^{\alpha2})}{2n_\square^{\alpha2}} - \ln \frac{(1-4b)}{4b}$$

$$\ln a_B = \ln a_{B,o} + \frac{1}{2}\ln \frac{(x_B - 2n_\square^{\alpha1})}{(x_B - 2n_\square^{\alpha2})} + \ln \frac{2n_\square^{\alpha2}(x_B - n_A^i)}{(x_B)^2}$$

$$- \frac{1}{2}\ln \frac{(1-4a)}{(1-4b)} - \ln \{4b\,[1-2(a+b)]\}$$

where $a_{A,o}$ and $a_{B,o}$ are the values of the activity of A or B, respectively, at the stoichiometric 1 : 1 composition. The defect numbers are now given as defect concentrations, i.e.

$$n_A^i = (N_A^i/N_L);\ n_\square^{\alpha1} = (N_\square^{\alpha1}/N_L);\ n_\square^{\alpha2} = (N_\square^{\alpha1}/N_L)$$

Usually the theoretical curve for $\ln a_B$ is fitted to the experimental activity values to obtain the parameters $a$ and $b$, and also $\ln a_{B,o}$.

If we now apply the equation to the data for the Te-activity in the Ni–Te system,[1] for example, we will discover that the theoretical curve reflects very well the inflection point in the composition dependence of $\ln a_{Te}$ but that we can never obtain perfect agreement with the experimental data by only changing the two parameters $a$ and $b$.

As pointed out earlier, we assumed that the different Gibbs energy values $\mu_A^i$, $\mu_\square^{a1}$, and $\mu_\square^{a2}$ are independent of composition, i.e. the Gibbs energy effect to remove an A-atom from the, say, α1-sublattice (and to form a vacancy, consequently) is the same, whether we are at a composition close to stoichiometry, and we therefore have only a few vacancies, or whether we are far away from stoichiometry at high concentrations of B, where we also have a large number of vacancies already present. Wagner and Schottky stated quite clearly that their treatment was applicable only in the vicinity of the stoichiometric composition, but many of the NiAs-phases exist only in composition ranges far away from it. A solution was offered by Anderson[3] who introduced additional parameters into the equations, i.e. pairwise interaction energies (or enthalpies) between similar defects: $H_{\square\square}$ and $H_{ii}$ in our case. They take care of the mutual repulsion (or attraction) between these defects, and their contribution becomes significant when the number of defects (be it vacancies or interstitial atoms) increases. However, we still assume a *statistical* distribution of these defects ('Zeroth Approximation' or 'Bragg–Williams Approximation').

In the NiAs-phases one has some problems with the counting of the number of nearest neighbour interactions on the same sublattice, mostly because the $c/a$-ratio can vary considerably, thereby changing the relative length of the distances between different neighbours. Some authors used a coordination number of $2 + 6 = 8$ for the vacancies at a distance $c/2$ above and below, and six others at a distance $a$ within the same layer. We used a similar approach, and to obtain the number of possible interactions between 'nearest neighbour' vacancies we assumed additionally that

$$H_{\square\square}^{inter} \approx 3 \times H_{\square\square}^{inter} = 3 \times H_{\square\square},$$

since $c/2$ is usually considerably shorter than $a$ and these interaction enthalpies will probably decrease with increasing distance $r$, i.e. with $1/r^n$ ($n \approx 2$?). For the interstitial A-atoms (in the trigonal-bipyramidal holes) the situation is much simpler: each $A_i$-atom can have 12 other $A_i$-atoms at the same distance as its nearest neighbours on the $i$-sublattice.

With these additional interaction enthalpies our equation for the total Gibbs energy now takes the form

$$G = (1/N_L) \; [N^t \mu^* + N_A{}^i \mu_A{}^i + N_\square^{\alpha 1} \mu_\square^{\alpha 1} = N_\square^{\alpha 2} \mu_\square^{\alpha 2} +$$

$$+ \quad \frac{12 \, (N_A{}^i)^2}{N^t} \; H_{ii} + \frac{12 \, (N_\square^{\alpha 1} + N_\square^{\alpha 2})}{N^t} \; H_{\square\square} \; ] \; - \; TS_{conf}$$

From this we can again derive equations for the activities of the components A and B; they look somewhat more complicated because they also contain now, of course, terms with these interaction enthalpies:

$$\ln a_A = \ln a_{A,\,o} + \ln \; \frac{(x_B - 2n_\square^{\alpha 2})}{2n_\square^{\alpha 2}} \; - \ln \frac{(1-4b)}{4b} \; -$$

$$\frac{(n_\square^{\alpha 1} + n_\square^{\alpha 2})}{x_B} \; \frac{12 H_{\square\square}}{RT} \; + \; (a+b) \; \frac{24 H_{\square\square}}{RT}$$

$$\ln a_B = \ln a_{B,\,o} + \frac{1}{2} \; \ln \; \frac{(x_B - 2n_\square^{\alpha 1})}{(x_B - 2n_\square^{\alpha 2})} \; + \ln \; \frac{2n_\square^{\alpha 2} \, (x_B - n_A{}^i)}{(x_B)^2} \; -$$

$$\frac{1}{2} \; \ln \; \frac{(1-4a)}{(1-4b)} \; - \; \ln \{4b \, [1-2 \, (a+b) \, ] \} +$$

$$+ \; \frac{(n_\square^{\alpha 1} + n_\square^{\alpha 2})}{x_B} \; \frac{12 H_{\square\square}}{RT} \; - \; \left( \frac{n_A{}^i}{x_B} \right)^2 \; \frac{6 H_{ii}}{RT} \; - \; \left[ \frac{(n_\square^{\alpha 1} + n_\square^{\alpha 2})}{x_B} \right]^2 \; \frac{6 H_{\square\square}}{RT} \; -$$

$$(a+b) \; \frac{24 H_{\square\square}}{RT} \; + (a+b)^2 \; \frac{24 \, (H_{ii} + H_{\square\square})}{RT}$$

The corresponding theoretical curve can now be fitted to the experimental tellurium activities in the Ni–Te system (at 873 K). The best agreement was obtained with the parameters $a = 0.01$, $b = 0.000001$, $H_{\square\square} = 2000$ J mol$^{-1}$, and ln $a_{Te, o} = -6.70$ ($H_{ii} = 0$). It has to be pointed out that the absolute values of the disorder parameters $a$ and $b$ are not very meaningful in this case since they refer to the disorder at the hypothetical stoichiometric (1 : 1)–composition where the phase does not exist. Really significant is only the large difference between $a$ and $b$ of several orders of magnitude which is typical for a partially filled CdI$_2$-type structure. The interaction energy (enthalpy) $H_{\square\square} = 2000$ J mol$^{-1}$ describes, at least formally, a slight repulsion between the vacancies in the nickel sublattice, and it compares reasonably well with $\Delta_{VVM}G = 3800$ J mol$^{-1}$ as obtained by Stølen and Grønvold for the same temperature.[4] Since the concentration of interstitial nickel atoms (on trigonal-bipyramidal sites) will be negligible on the tellurium-rich side of stoichiometry, $H_{ii}$ was taken to be zero to simplify the calculation of the curve.

There exists experimental evidence in the Ni–Te system for a second order transition from the NiAs-type to the CdI$_2$-type structure, probably somewhere between 55 and 57 at% Te.[5,6] Since, however, the model curves for both structure types are practically identical in this composition range relatively close to the (1 : 1)-stoichiometry we preferred to use one single curve, based on nickel vacancies in every other transition metal layer, for the entire homogeneity range of the phase.

## Application of the Model

The next few examples should demonstrate the application of our model to several other binary systems of transition metals with chalcogens or with antimony. For all these systems, with one exception (Fe–S), partial molar thermodynamic properties were measured in our laboratory using the isopiestic method originally conceived by Herasymenko.[7] For most of the systems it was also necessary to clarify the corresponding phase relationships.

*The Fe–Te system*

Application of our model to the activity measurements[8] results in a curve with $a = 0.01$, $b = 0.000001$, $H_{\square\square} = 1000$ J mol$^{-1}$, and ln $a_{Te, o} = -5.09$ if we draw one single theoretical curve through the data points of both phases with NiAs-related structures, monoclinic $\partial$ and hexagonal $\partial'$. We think that the slight horizontal inflection point in the composition dependence of the activities is real, and that this proves a CdI$_2$-type vacancy distribution, i.e. that the vacancies are more or less restricted to alternate layers.

From a Gibbs–Duhem integration we could obtain the integral Gibbs energies of mixing for these two phases. If we compare the experimental $\Delta$ G-values with those calculated from the model, the agreement is very good. This shows that in principle the model is also able to describe integral Gibbs energy curves, but the partial Gibbs energies, i.e. the thermodynamic activities, are much more sensitive to variations of the model parameters.

*The Fe–Se system*

In the Fe–Se system[9] we have at 873 K the hexagonal $Fe_{1-x}$Se-phase which changes apparently continuously into a monoclinic structure around 55 at% Se, and – separated by a two-phase field – the monoclinic $Fe_3Se_4$-phase with an ordered vacancy distribution of the $Cr_3S_4$-type. A great variety of additional superstructures exists at lower temperatures, and around $Fe_7Se_8$ one can observe the gradual disordering of the vacancies with increasing temperature. From the crystallographic studies it appears that at 873 K the vacancies are distributed at random over all iron layers in this $Fe_{1-x}$Se-phase.

The theoretical curve with $a = b = 0.0025$ (indicating an equal vacancy concentration in all iron layers), with $H_{\square\square} = 21\,000$ J mol$^{-1}$, and ln $a_{Se,\,o} = -11.30$, describes the experimental data very well. One can see that $a$ (or better: $a + b$) has become smaller compared to the tellurium systems (Ni–Te and Fe–Te) pointing to a smaller degree of thermal disorder at the stoichiometric composition. Since the iron-rich boundary of $Fe_{1-x}$Se is much closer to the stoichiometric composition than for the corresponding phase in the Fe–Te system, the value of $a$ (and $b$) becomes much more significant. One can also see that the interaction energy is much larger than that for the tellurium systems, i.e. there is a much stronger repulsion between the iron vacancies in the Fe–Se system. Our value of $H_{\square\square} = 21\,000$ J mol$^{-1}$ is again in reasonable agreement with the value of $\Delta G_{VVM} = 29\,600$ J mol$^{-1}$ obtained by Stølen and Grønvold for 873 K.[4]

For the $Fe_3Se_4$–phase we calculated a theoretical curve using a different model, i.e. a model based on a $Cr_3S_4$-type superstructure, where the vacancies are restricted to every other layer, and we have 'strings' of transition metal atoms alternating with 'strings' of vacancies in these layers. For the model calculations we assumed that the deviation from the (3 : 4)-stoichiometry is achieved by additional vacancies on the selenium-rich side and by a gradual filling of vacant positions on the iron-rich side. The corresponding curve was obtained with a disorder parameter $a_{(3:4)} = 0.004$ (where $a_{(3:4)} = (N^b/N^t)_{(3:4)}$) which means that 3.2% of the iron atoms in alternate iron layers change place with neighbouring vacancies at 873 K due to thermal agitation.

*The Fe–S system*

The experimental sulphur activities for the pyrrhotite phase in the Fe–S system were obtained by an $H_2/H_2S$ equilibration method.[10] The application of our model to this particular set of data yields a disorder parameter $a = b = 0.0018$ (again smaller than in the Fe–Se system), with $H_{\square\square} = 40\,000$ J mol$^{-1}$ and ln $a_{s,o} = -9.72$. This value of the interaction energy compares with $\Delta_{VVM}G = 53\,500$ J mol$^{-1}$ at 1073 K from Stølen and Grønvold, obtained from several sets of data by different authors.[4]

*The Pd–Te system*

The Pd–Te system is of special interest since the PdTe-phase with NiAs-structure is separated from the $PdTe_2$-phase with $CdI_2$-structure by a two-phase field. Although older phase diagrams showed at higher temperatures a continuous solid solution between these two phases, we are certain from thermodynamic and phase diagram studies that they are separate phases at all temperatures.[11]

The application of our model to the tellurium activities in the PdTe-phase yields $a = b = 0.005$, $H_{\square\square} = 1600$ J mol$^{-1}$, and $\ln a_{Te,o} = -4.15$, and the data in the PdTe$_2$-phase can be reasonably well described by a curve with $a = 0.01$, $b = 0.000001$ and $\ln a_{Te,o} = -6.24$. Due to the difficulties in obtaining reliable $\Delta H_{Te}$-data for the PdTe$_2$-phase from the isopiestic measurements it is quite possible that the experimental activity data carry a relatively large error which might be the reason for a deviation of the model curve from the data points.

The value of $H_{\square\square} = 1600$ J mol$^{-1}$ for PdTe is of a similar magnitude as in other tellurium systems; for PdTe$_2$ it is impossible to estimate any value for $H_{\square\square}$ since the phase is too narrow.

*The Co–Sb system*

In the Co–Sb system, where the homogeneity range of the NiAs-phase extends to both sides of the stoichiometric composition, we have cobalt vacancies on the antimony-rich side and additional cobalt atoms in the trigonal-bipyramidal positions on the cobalt-rich side.[12] Our model can be applied equally well on both sides of stoichiometry, yielding one continuous theoretical activity curve. The disorder parameter $a = b = 0.005$ means that at the stoichiometric composition, at 1173 K, 2% of the cobalt atoms leave their regular octahedral sites and enter interstitial (trigonal–bipyramidal) positions due to thermal disorder. In this system we have to consider also interactions between these interstitial cobalt atoms besides those between the vacancies. Both interaction energy terms are negative ($H_{\square\square} = -600$ J mol$^{-1}$; $H_{ii} = -1700$ J mol$^{-1}$), which would indicate that like defects attract each other.

## Conclusions

Although much remains to be done in these systems with NiAs-phases, some conclusions – for the sake of simplicity restricted to the eighth group elements – can be drawn from our thermodynamic studies. For the chalcogen systems the modelling has shown that a sufficiently correct description of the change of the partial molar properties across the homogeneity range of the NiAs-phases can only be obtained if one assumes an interaction between the vacancies (and also between the interstitial transition metal atoms). These interaction energy terms may contain some other secondary factors (the model is rather too simple), but in general one can safely assume that the magnitude of the interaction energy does reflect the extent – or better is a measure – of the mutual repulsion or attraction of the vacancies.

In all chalcogen systems the values of the interaction energy per vacancy pair are positive, i.e. we have a repulsion between the vacancies, and the values increase from the tellurides to the selenides, and are highest for the sulphides while the increase from Ni to Co and finally to Fe is not as clearly noticeable but the trend is there.

With a repulsion between the vacancies one would expect the following:

(1) With an increase of the interaction energy the tendency for the formation of superstructures should increase, i.e. the random distribution of vacancies will be replaced by the more stable ordered distribution. This is borne out by experiments: while in Ni–Te and Co–Te so far no superstructures have been observed, they have been definitely established in the Ni–Se and Co–Se system, and in Ni–S and Co–S the NiAs-phases occur only at high temperatures, not to mention the various superstructures in pyrrhotite (Fe–S).

(2) With increasing repulsion between vacancies one would expect a shift of the chalcogen-rich phase boundaries towards smaller vacancy concentrations, i.e. the phase regions should become narrower. This is again in accord with the experimental evidence: While in Ni–Te the phase boundary extends to 66.6 at% Te, the chalcogen-rich limits in Co–Se and Ni–Se are 59 and 57 at% Se, respectively and still less in the sulphide systems.

The reason for the increase in interaction energies might be attributed to the increase in ionic character. Since in NiAs-phases a low $c/a$-ratio is characterised by metallic, a high $c/a$-ratio by ionic bonding, a correlation should also exist between the interaction energies and the $c/a$-ratios. This is indeed the case: the interaction energy increases (and also the enthalpy of formation) with increasing $c/a$-ratio and the assumption that the increase of repulsion is related to the increase in ionic bonding seems to be plausible. It is interesting to find that even the small negative value for the system Co-Sb fits quite well into this scheme, a negative interaction energy possibly associated with metallic behaviour.

# References

1. M. Ettenberg, K. L. Komarek, and E. Miller: 'Thermodynamic Properties of Nickel-Tellurium Alloys,' *J. Solid State chem.*, 1970, **1** (3–4), 583–592.
2. C. Wagner and W. Schottky: 'Theorie der geordneten Mischphasen,' *Z. Physik. Chem.*, 1931, **B11** (2/3), 163–210.
3. J. S. Anderson: 'The Conditions of Equilibrium of Non-stoichiometric Chemical Compounds,' *Proc. Roy. Soc. London*, 1946, **185A**, 69–89.
4. S. Stølen and F. Grønvold: 'Thermodynamics of Point Defects in Monochalcogenides of Iron, Cobalt and Nickel with NiAs-Type Structure,' *J. Phys. Chem. Solids*, 1987, **48** (12), 1213–1225.
5. R. S. Carbonara and M. Hoch: 'Thermodynamik und Struktur von $Ni_{1-x}$ Te,' *Monatsh. Chem.*, 1972, **103** (3), 695–715.
6. P. Coffin, A. J. Jacobson and B. E. F. Fender: 'The Observation of an Order-Disorder Transition in the Ni-Te System by Neutron Diffraction,' *J. Phys. C: Solid State Phys.*, 1974, **7** (16), 2781–2790.
7. P. Herasymenko: 'Thermodynamic Properties of Solid Solutions. Vapor Pressures of Cadmium over Alpha Silver-Cadmium Alloys,' *Acta Met.*, 1956, **4** (1), 1–6.

8. H. Ipser and K. L. Komarek: 'Thermodynamic Properties of Iron-Tellurium Alloys,' *Monatsh. Chem.* 1974, **105** (6), 1344–1361.

9. W. Schuster, H. Ipser and K. L. Komarek: 'Thermodynamic Properties of Iron-Selenium Alloys,' *Monatsh. Chem.*, 1979, **110** (5), 1171–1188.

10. R. Y. Lin, H. Ipser and Y. A. Chang: 'Activity of Sulfur in Pyrrhotite at 1073 K,' *Met. Trans.*, 1977, **8B** (2), 345–346.

11. H. Ipser: 'Thermodynamic Properties of Palladium-Tellurium Alloys,' *Z. Metallkde*, 1982, **73** (3), 151–158.

12. G. Hanninger, H. Ipser and K.L. Komarek: 'Thermodynamic Properties of NiAs-Type $Co_{1+x}Sb$,' *Z. Metallkde*, 1990, **81** (5), 330–334.

# Some Anecdotes about the Production of Carbides and Nitrides

B. S. TERRY AND P. GRIEVESON

*Department of Materials, Imperial College, London, UK*

**Abstract**

An attempt is made to discuss some interesting observations made in the preparation of carbides and nitrides. Some consideration is given to the formation of silicon carbide whiskers and proposals are made for the best conditions of production. Observations are made on the preparation of silicon nitride from silicon powder in nitrogen and hydrogen gas mixtures. Anomalies relating to the variation of reaction rate with nitrogen partial pressure are explained in terms of the participation of silicon monoxide as an intermediary. Anomalies on the formation of iron and titanium carbide from ilmenite by carbothermic reduction are explained in terms of interfacial phenomena. In the preparation of tungsten carbide from tungsten trioxide, the apparent anomalous variation of the reaction rates with temperature are explained by a variation in the reaction mechanisms.

## Preparation of Silicon Carbide Whiskers

Silicon carbide whiskers and fibres have recently attracted enormous interest because of their useful mechanical properties. A systematic study of the formation of silicon carbide whiskers has been carried out by reducing different forms of silica with carbon under different partial pressures of CO gas. Of prime interest are the effect of temperature and the role of silicon monoxide gas pressure.[1]

Three different forms of silica were used as starting materials to investigate the effect of topology and surface area on the reaction rate. These were crushed quartz (99.98% pure) precipitated silica (99.98%) and crushed gel silica glass (99.9%).

Appropriate amounts of silica and lamp black, as carbon source, were weighed so that twice as much carbon as that required to produce silicon carbide by stoichiometric reaction was used. In some cases, 5 wt % $FeCl_3$ was added to study its role as a catalyst. The pelletised specimens were held in a graphite crucible and heated using a radio frequency induction coil. A Cartesian manostat was used to control the total pressure in the reaction vessel. Excess gas produced by the reaction was released through the manostat and collected and its volume measured. Samples were studied by X-ray diffraction analysis using a Gurnier focusing camera. Reaction products were also examined using scanning and transmission electron microscopy.

The rate of reduction curves for quartz and precipitated silica are shown in Figs 1 and 2 at various CO isobars. Both the reduction rate and extent of reaction

111

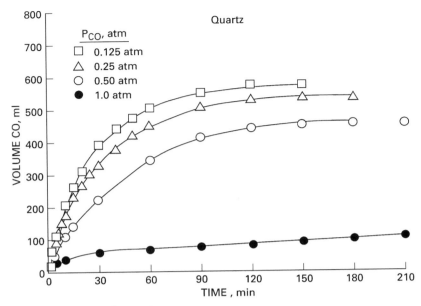

**Fig. 1** *Reduction curve for quartz.*

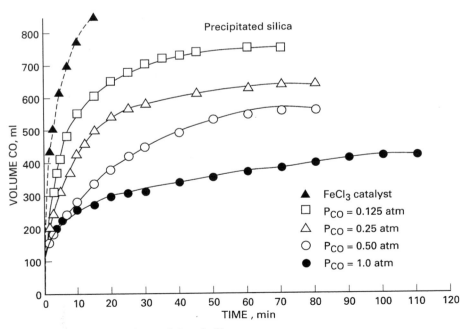

**Fig. 2** *Reduction curve for precipitated silica.*

increase with temperature. Similar behaviour is observed when the total pressure is reduced as expected from Le Chatelier's principle. A limited number of experiments was carried out with $FeCl_3$ additions as a catalyst for the carbon oxidation reaction. Its presence increased the rate of reaction significantly as shown in Fig. 2.

Verification of the formation of SiC whiskers was obtained by scanning electron microscopic examination. The average volume fraction of whiskers in a given sample was determined by quantitative optical microscopy and statistical analysis. There were experimental conditions where significant reduction of silica produced no whiskers as in the cases of reduction in the presence of $FeCl_3$. In general terms, the factors controlling the formation of whiskers are summarised as

(i)  whiskers do not form at high temperatures and low CO partial pressures, and
(ii)  the volume fraction of whiskers significantly drops at high CO pressure and low temperature.

These observations indicate that an intermediate temperature and pressure range are essential for a high product yield.

It was apparent that two overall reactions are important in the production of silicon carbide whiskers, viz: the silicon monoxide formation reaction:

$$SiO_2(s) + C(s) = SiO(g) + CO(g) \qquad (1)$$

and the silicon monoxide utilisation reaction:

$$SiO(g) + C(s) = SiC(s) + CO(g) \qquad (2)$$

These equilibria and their variation with temperature are represented in Fig. 3. Also included in Fig. 3 are the conditions for optimum silicon carbide whisker formation as shown in the shaded area.

## Silicon Nitride Preparation from Silicon Powder and Nitrogen–Hydrogen Gas Mixtures

It is well known that in a reaction between a gas and a solid to produce a solid product, as in the case of silicon with nitrogen:

$$3Si(g) + 2N_2(g) = Si_3N_4(s) \qquad (3)$$

the reaction rate is expected to increase with increasing gas partial pressure.

In a preliminary series of experiments the effect of nitrogen partial pressure on the reaction with silicon was studied at 1300°C for a period of 24 h. Hydrogen was chosen as a diluent to prevent the oxidation of silicon to silica. The results are shown in Fig. 4. It is seen that the maximum rate is achieved at $p_{N_2} = 0.8$ bar and not in

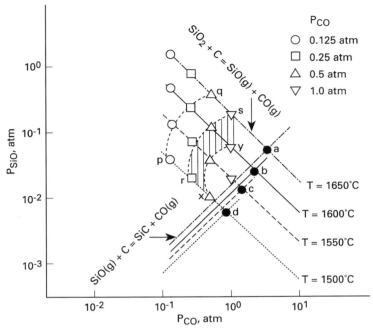

**Fig. 3** *Equilibria to SiO and SiC fraction.*

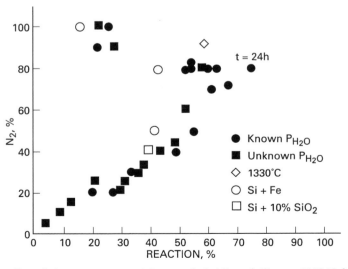

**Fig. 4** *The effect of nitrogen pressure of the rate of nitriding of silicas at 1300°C for 24 h.*

pure nitrogen as expected. It seemed that hydrogen played a significant role and it was considered that this role may be in establishing an oxygen potential necessary for the formation of silicon nitride via silicon monoxide formation. To test this hypothesis, all subsequent experiments were carried out by equilibrating the gas with $Cr_2O_3$-Cr at various temperatures to produce known partial pressures of water vapour.

A series of time experiments was carried out with a gas mixture containing 80% $N_2$ and 20% $H_2$ with a variety of particle sizes for the silicon specimen and known partial pressures of water vapour. Some typical results are shown for the reaction extent with time in Fig. 5. As expected, the extent of reaction varied with particle size. An analysis of the data for a chemical reaction controlled regime with a reaction product at the surface suggests that a plot of $1-(1-X)^{\frac{1}{3}}$ against time/diameter of particle should be linear, if $X$ is the fraction of reaction which has occurred. The data for all experiments were tested for this mechanism.

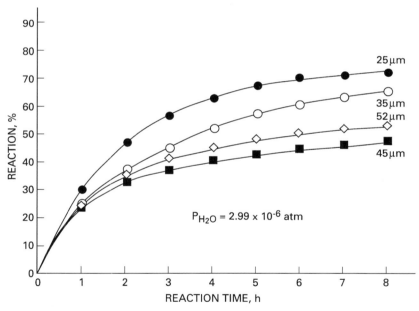

**Fig. 5** *Extent of nitriding reaction with time at 1300 °C.*

The data for all particle sizes with the exception of 3 μm and 1.5 μm particles for a fixed nitrogen pressure gave identical linear plots as seen for 45 μm particle silicon powder in Fig. 6. Surprisingly, the chemical reaction rate constant was the same for all water vapour partial pressures used but was found to be dependent on the partial pressure of nitrogen. Later experiments showed that under these conditions, gas starvation was the reason for non agreement with the fine particle sized material.

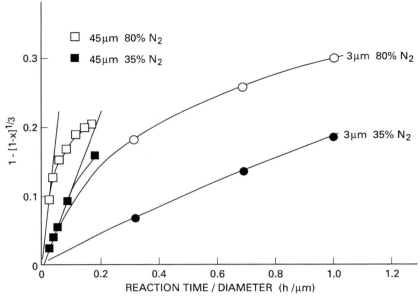

**Fig. 6** *Chemical reaction control test for nitriding silicon.*

These results indicated that the reaction controlling step varied with time and a study of the late stages of reaction indicated that this was pore diffusion controlled. A series of density measurements and porosity determinations confirmed this conclusion. Experimental results showed that the rate of the pore diffusion controlled regime was directly proportional to the partial pressure of water vapour in the gas. It was thus apparent that the later stages of reaction were controlled by pore diffusion of water vapour through the reacted silicon nitride zone at the specimen surface with diffusion coefficients of $\sim 10^{-2} cm^2/s$.

A further surprising observation was made when the specimens were examined by X-ray diffraction. It was noted that during the chemical reaction controlled regime only $\alpha$-silicon nitride was formed. Later, increasing proportions of $\beta$-silicon nitride were formed when water vapour pore diffusion was controlling as seen in Fig. 7.

Thus in the early stages of nitriding, $\alpha$-$Si_3N_4$ is formed by a chemical reaction controlled mechanism which is dependent on nitrogen pressure. At the onset of water vapour pore diffusion control, it seems that only $\beta$-$Si_3N_4$ is formed, giving a mixture of $\alpha$- and $\beta$-$Si_3N_4$ in the final product. These results indicate that it is possible to produce pure $\alpha$-$Si_3N_4$ by carefully increasing the water vapour partial pressure in the system as reaction proceeds.

## Carburisation-Reduction of Ilmenite Ores

The paint pigment industry has long been interested in producing titania pigment from ilmenite rather than rutile because it is plentiful and cheap. All efforts to cause

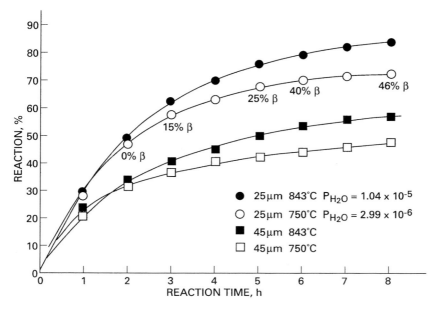

**Fig. 7** *The proportion of α- and β- Si₃N₄ variation with reaction time.*

separation of iron and reduced rutile after hydrogen or carbothermic reduction at low temperatures have failed because of the finely dispersed morphology of the iron in the reduced material. It was considered that higher temperature reaction with carbon to form iron and titanium carbide might be advantageous because

(i)   iron will absorb carbon and melt with agglomeration and separation from titanium carbide;
(ii)  any manganese impurity will associate with the iron; and
(iii) titanium carbide chlorinates more efficiently than rutile.

With this in mind we set out to study the carburisation reduction of ilmenite with coal in the temperature range 1314–1517°C². At all temperature the initial reaction is extremely rapid as shown in Figs 8 and 9. At temperatures below ~1415°C, it was noted that titanium oxycarbide was formed from $Ti_3O_5$ (see Table 1), whilst at higher temperatures, the oxycarbide formed from $Ti_2O_3$.

Thus, the reaction sequence can be divided into two regions:

1. Very fast reaction to produce Fe and $Ti_3O_5$ or $Ti_2O_3$.
2. Slower reaction to produce titanium oxycarbide of particular interest was the observation of where the titanium oxycarbide formed. After about 40 min of 1314°C, the formation of the oxycarbide phase was observed micrographically. It was seen that titanium oxycarbide always formed associated with iron and never observed to nucleate at the $Ti_3O_5$ gas interface. At this stage, we were

117

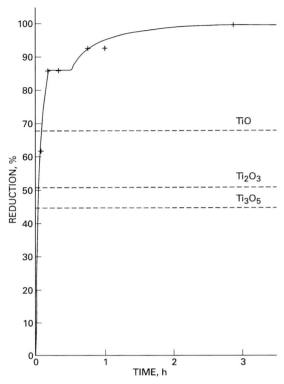

**Fig. 8** *WMS ilmenite + coal at 1517 °C: % reduction v. time.*

**Table 1** *X-ray diffraction analysis for WMS ilmenite reduce with collie coal 1413 °C.*

| Reduction Time | Reduction % | Phase present |
|:---:|:---:|:---|
| 5 min | 46 | $M_3O_5$, Fe, $TiO_xC_y$ |
| 10 min | 51 | $M_3O_5$, Fe, $TiO_xC_y$ |
| 15 min | 64 | $M_3O_5$, Fe, $TiO_xC_y$ |
| 30 min | 80 | $M_3O_5$, Fe, $TiO_xC_y$ |
| 1 h | 87 | $M_3O_5$ (trace), Fe, $TiO_xC_y$ |
| 1 ¾ h | 90 | Fe*, $TiO_xC_y$, $Fe_3C$ (trace) |
| 4 h | 90 | $TiO_xC_y$, Fe,* $Fe_3C$ (trace) |
| 20 h | 97 | $TiO_xC_y$, Fe |

* Iron in this case appears to be slightly martensitic.

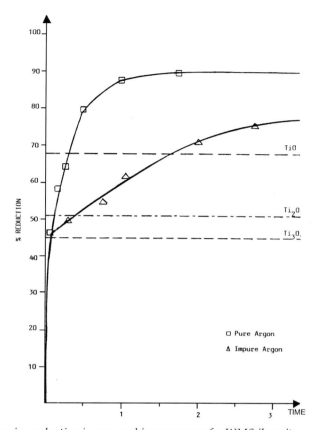

**Fig. 9** *Comparing reduction in pure and impure argon for WMS ilmenite + coal at 1413°C.*

delighted to note that titanium oxycarbide existed as a separate phase between iron and titanium oxide and clear separation from the iron phase. However, as the reaction proceeds it was noticed that the oxycarbide phase became dispersed in the iron. This observation is the exact opposite of the well known cases of Ostwald ripening. If we look at the factors important in Ostwald ripening, we see that the only important variable is the interfacial tension.

It is well known that a high metal surface tension is ideal for separation. It is also known that the presence of transition metal and carbon together in liquid iron can give associative adsorption with a lowering of the surface tension, since one of the factors that is likely to occur is an increase of titanium and carbon in the iron with increased reaction time as the overall oxygen potential in the system is reduced. Thus associative adsorption seems a likely answer to the later dispersion of the oxycarbide in the liquid metal. Unfortunately, the dispersion of oxycarbide in iron defeats one of the major objectives of the study, i.e. easy separation of the two phases. However, the dispersion of refractory materials in liquid metals is of prime

119

interest in the production of composite materials. Recent work has shown that wear resistant composites of TiC in iron can be produced by carburisation-reduction of ilmenite.

## Tungsten Carbide Preparation from Tungsten Oxide

It is well established that reaction rates increase with increasing temperature. It was surprising, therefore, when we carried out some preliminary experiments on the carburisation reduction of $WO_3$ in carbon monoxide that the time to convert the oxide to WC was the same at 800°C and 1120°C, whilst a longer reaction time was necessary at 1025°C.

These results are only explicable if we have a change in reaction mechanism. Further experiments at various times indeed showed this to be the case (see Fig. 10). It can be seen that at 800°C the reaction proceeds with a weight loss at all times, whilst at higher temperatures an initial weight loss is followed by a weight gain. The results indicate there are two main categories of reaction:

1. reduction-carburisation of tungsten oxide at temperatures below about 900°C;
2. the reduction to tungsten and carburisation of tungsten at temperatures above about 900°C.

At temperatures below 900°C, the reduction-carburisation proceeded very rapidly to produce WC. The results of time experiments showed that $WO_3$ was reduced initially to $WO_2$. After almost complete reduction to $WO_2$, WC formation is observed. The initial oxide reduce rate is linear with time, whilst after the nucleation of WC on $WO_2$ the rate becomes parabolic. At 800°C, complete conversion to WC occurs in 2 h. It is also interesting to note that the WC formed at 800°C has a particle size of ~0.15µm.

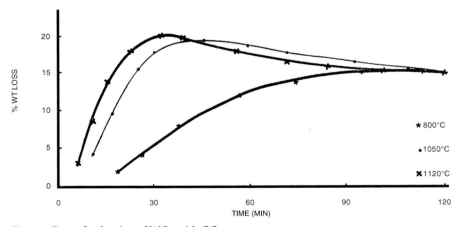

**Fig. 10** *Rate of reduction of $WO_3$ with CO.*

In contrast, at the higher temperatures, the reaction proceeds by reduction of the oxide to tungsten metal, which takes 40 min at 1050°C and 30 min at 1120°C. When carburisation occurs, $W_2C$ is formed initially and grows until ~ $\frac{1}{3}$ of the tungsten is converted. At this stage, WC nucleates on the surface of the $W_2C$ after ~25 min at 1125°C and 30 min at 1050°C. WC continues to grow on the $W_2C$ until all the $W_2C$ is consumed, whereupon WC continues to grow on the tungsten until reaction is complete.

The kinetics of the reaction were studied using tungsten powder of 2μm particle size and shown to be diffusion controlled. Since initially $W_2C$ grows on W and finally WC grows on W, it is possible to obtain information on the permeability of both $W_2C$ and WC. This information is presented in Fig. 11.

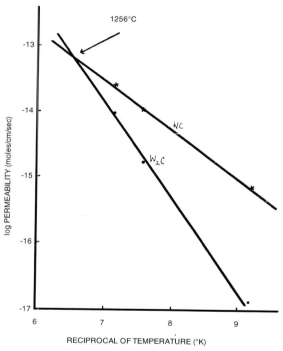

**Fig. 11** *Relationship between log permeability vs temperature for WC and $W_2C$.*

As expected from the experimental observations, the permeability of carbon through WC was observed to be greater than that through $W_2C$ at all temperatures studied. This was anticipated because once WC nucleates it quickly converts all the $W_2C$. Extrapolation of the permeability values indicates that they become equal at 1256°C. This value is in excellent agreement with the temperature at 1250°C which is accepted as the $W_2C$ eutectoid temperature.

121

**Table 2** *X-ray diffraction analysis of WMs ilmenite reduced with collie coal at 1517°C.*

| Time of Reduction | Reduction % | Phases present |
|---|---|---|
| 5 min | 62 | $M_2O_3$, $TiO_xC_y$ Fe, $M_3O_5$, (Trace) |
| 10 min | 87 | $M_2O_3$, $TiO_xC_y$ Fe |
| 18 min | 87 | $M_2O_3$, $TiO_xC_y$, Fe C |
| 30 min | 87 | $M_2O_3$, $TiO_xC_y$, Fe, C ($Fe_3C$) (Trace) |
| 45 min | 93 | $TiO_xC_y$, Fe, C |
| 1 h | 93 | $TiO_xC_y$, Fe C ($Fe_3C$) (Trace) |
| 3 h, 8 min | 100 | $TiO_xC_y$, Fe, C |

## Acknowledgements

The financial support of SERC, Joseph Lucas Industries, Tioxide International and B.T.G. are gratefully acknowledged. We also wish to thank Dr A. Jha, Dr A. Chrysanthou, Dr K. C. Coley and Dr S. P. Rathband whose work formed the basis of this paper.

## References

1. A. Chrysanthou, P. Grieveson and A. Jha: 'Formation of Silicon Carbide Whiskers and their Microstructure,' *J. Mat. Science*, 1991, **26**, 3463–3476.
2. K. S. Coley, B. S. Terry and P. Grieveson: 'Carburisation-Reduction of Ilminite Ores,' *Proceedings of Terkel Rosenqvist Symposium*, Norwegian Institute of Technology, Trondheim, May 1988, pp151–176.

# Self-Propagating High Temperature Synthesis of $Al_2O_3$–TiC and $Al_2O_3$–TiB$_2$ Composites

A. Chrysanthou

*Department of Materials Engineering and Materials Design, University of Nottingham, Nottingham NG7 2RD, UK*

### Abstract

Self-propagating high temperature synthesis was used as a means to prepare $Al_2O_3$–TiC and $Al_2O_3$–TiB$_2$ composites. The ignition and adiabatic temperatures depended on the reactant composition as well as the amount of precompaction. The effect of these parameters on the product were examined. The microstructures of the products from different reactions were compared.

## Introduction

Self-propagating high temperature synthesis (SHS) involves the ignition of an exothermic reaction whereby the evolution of heat is sufficient to sustain the reaction. The enthalpy of a reaction may be calculated using the equation

$$\Delta H°_{T_{ig}} = \int_{T_{ig}}^{T_{ad}} \Delta C_p \, dT$$

where, $\Delta H°_{T_{ig}}$ is the enthalpy change at the ignition temperature, $\Delta C_p$ is the change in heat capacity, $T_{ig}$ is the ignition temperature and $T_{ad}$ is the adiabatic (also referred to as combustion) temperature. Where changes of state are involved appropriate changes in the equation are required. It has been suggested that for a reaction to be self-sustaining, the ratio $\Delta H°/\Delta C_p$ must be in excess of 2000.[1] The reaction occurs in two different modes.[2] When it starts simultaneously in the whole sample, an explosion takes place. On the other hand, the reaction can ignite at one point in the sample and propagate in the form of a wave. The technique of SHS was first investigated in the 1950s.[3,4] During the next three decades researchers in the Soviet Union including Merzhanov and Borovinskaya[5,6] and Navozhilov[7] developed the SHS process further. Nowadays materials including ceramics, cermets, composites and intermetallics can be produced commercially.[8] There are various claims that materials produced by SHS have superior properties compared to those obtained by conventional methods[9] including an easier ability to sinter[10] and a stronger bond between components of composite materials. Demonstrated advantages include a low energy requirement, a fast production rate and a higher purity.[11]

During this study ceramic composite materials of $Al_2O_3$–TiC and $Al_2O_3$–TiB$_2$ were prepared by SHS. These materials are hard, strong and wear-resistant and have applications as cutting tools. Various SHS routes were investigated and the products compared. The first route can be expressed by the equation

$$Ti + C + xAl_2O_3 \rightarrow TiC + xAl_2O_3 \qquad (1)$$

where the amount of $Al_2O_3$ present was varied, thus producing a ceramic composite of different compositions. In addition, $Al_2O_3$ absorbed some of the exothermic heat and so moderated the reaction. A study of the reaction

$$3TiO_2 + 4Al + 3C \rightarrow 3TiC + 2Al_2O_3 \qquad (2)$$

was carried out for comparison. Products which were prepared from this route gave a fixed composition of 53.1% $Al_2O_3$ and 46.9% TiC. Composite materials obtained using SHS by reaction (2) have previously been prepared by Abramovici[12] who showed that the product can be sintered to 95% relative density. Adachi et al[13], who performed the same process under the application of mechanical pressure, reported products of superior fracture toughness compared to commercial ones. Requena et al.[14] who studied reaction (2) by means of differential thermal analysis (DTA) erroneously associated an endothermic reaction at 1000°C with the reduction of $TiO_2$ by aluminium, which is of course exothermic. The present study has shed light on the nature of the endothermic reaction at 1000°C. Other results which are in contrast with previous studies are also reported.

Some results are also reported for the system $Al_2O_3$-TiB$_2$ where the reaction

$$Ti + 2B + xAl_2O_3 \rightarrow TiB_2 + xAl_2O_3 \qquad (3)$$

was studied.

## Experimental

The raw materials used were Ti (< 45 μm particle diameter), B (< 45 μm particle diameter), $B_2O_3$, $Al_2O_3$ (in the form of corundum, varying from 20 μm to 150 μm particle diameter) and carbon black. Appropriate amounts of each reactant were weighed out and thoroughly mixed. Each mixture, which weighed a total of 20 g, was placed in a graphite crucible and patted down with a metal rod. The effect of compaction was studied by pressing samples at various pressures in the graphite crucible. All experiments were conducted in an induction furnace under static argon at atmospheric pressure. A constant heating rate of about 400°C/min was used throughout. Temperature measurement was carried out by means of a pyrometer connected to a computer for accurate data readings and storage. Reacted samples were allowed to cool in the furnace for 1–2 h. The products were characterised by X-ray diffraction (XRD) and optical microscopy.

## Results

A preliminary study was carried out to examine the effect of varying the Ti particle size on the reaction with carbon black. In both cases, for a Ti particle size in excess of 150 µm and a heating rate of 400°C/min, no self-propagating reactions were observed. The large particle size meant that Ti was acting as a diluent by absorbing much of the exothermic heat. For Ti between 45 µm and 150 µm, an ignition temperature of 1160°C was observed and 1096°C for Ti below 45 µm. All further work was carried out using Ti with a particle size below 45 µm.

*Reaction 1*

The effect of $Al_2O_3$ dilution on the reaction between Ti and carbon black is shown in Fig. 1. The ignition temperature increased with the $Al_2O_3$ content while the combustion temperature dropped. Upon ignition the sample temperature increased quite sharply and reactions took place in the form of an explosion. The time of reaction was considered to be the time taken to reach the combustion temperature once ignition occurred. Most reactions were complete in less than a second after ignition as shown in Fig. 2. In this plot the time at ignition was taken as zero time. The time of reaction was shorter for lower $Al_2O_3$ contents. As the $Al_2O_3$ content increased a number of extra temperature rises were observed after the maximum temperature had been reached. This is evident in Fig. 3 which shows the variation in temperature with time for a sample containing 20 wt% $Al_2O_3$ in contrast to Fig. 2 (for 10 wt% $Al_2O_3$).

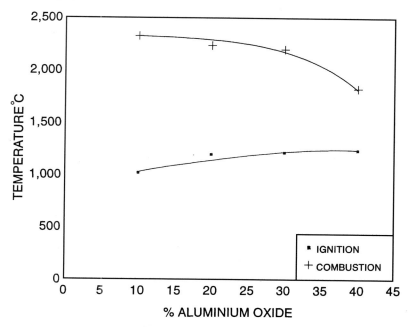

**Fig. 1.** *Combustion and ignition temperatures as a function of $Al_2O_3$ content.*

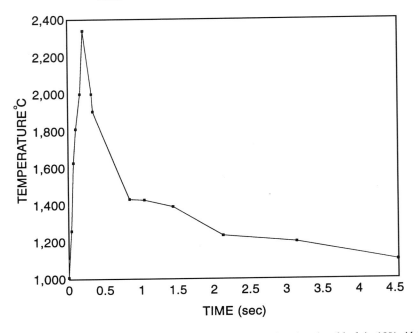

**Fig. 2.** *Temperature against time plot for reaction between Ti and carbon black in 10% Al$_2$O$_3$.*

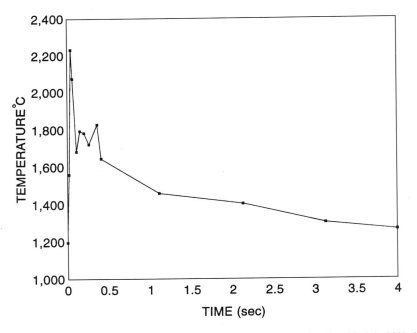

**Fig. 3.** *Variation of temperature with time for reaction between Ti and carbon black in 20% Al$_2$O$_3$.*

When the amount of $Al_2O_3$ present exceeded a critical level of about 41 wt% reaction was no longer self-propagating. However, if the reactant powders were not homogeneously mixed, this would result in regions rich in Ti and carbon black such that it was still possible for ignition to take place leading to a self-sustaining reaction. This is evident in Fig. 4 which shows the variation in temperature with time for a sample containing 53 wt% $Al_2O_3$ which reacted in an SHS mode.

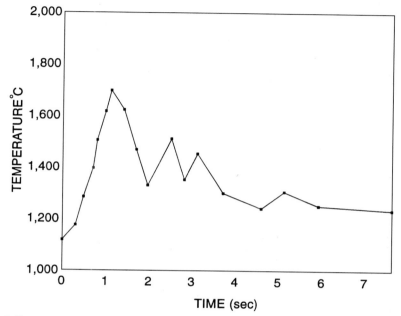

**Fig. 4.** *Temperature-time plot for reaction between Ti and carbon black in 53% $Al_2O_3$.*

On completion of the reaction the particles tended to bond together to produce large porous agglomerates. Figure 5 is a micrograph of a sample containing 35 wt% $Al_2O_3$ that reacted in a self-propagating manner. $Al_2O_3$ was wholly distributed within rivers of TiC particles with a significant amount of porosity present. The level of porosity was not affected by changing the starting sample composition and remained at about 40% throughout. For samples where no SHS took place there was no mixing between TiC and $Al_2O_3$ as shown in Fig. 6. Precompaction of samples seemed to decrease the ignition temperature. The precompaction of samples seemed to decrease the ignition temperature. However, for precompaction pressures above 2.8 MPa the reaction was no longer self-sustaining.

*Reaction 2*

For reaction 2 using stoichiometric amounts of TiO$_2$, Al and C, an ignition temperature of 1310°C was observed. The reaction did not take place in the explosion mode but as a wave front propagating through the sample. The reaction initiated

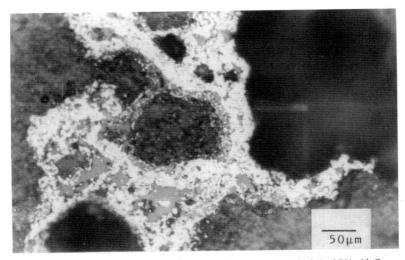

**Fig. 5.** *Microstructure from the reaction between Ti and carbon black in 35% Al$_2$O$_3$. black = Al$_2$O$_3$, white = TiC.*

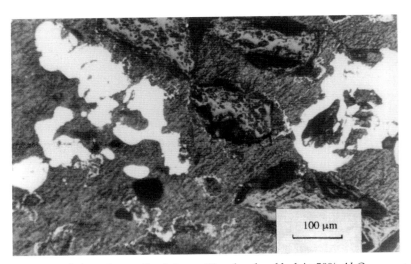

**Fig. 6.** *Micrograph from the reaction between Ti and carbon black in 50% Al$_2$O$_3$. black = Al$_2$O$_3$, white = TiC.*

at the sample edge and it took about 4 s for the wave to move through the sample. Evolution of CO gas was observed just prior to ignition and this was accompanied with a weight loss of about 5%. XRD analysis of samples just prior to ignition showed that a small amount of TiO$_2$ reacted with carbon black. During reaction, the temperature increased in two bursts. The first one was a temperature rise to 1700°C after about 1 s followed by a second burst to about 2100°C. XRD analysis confirmed the production of TiC and Al$_2$O$_3$ but also indicated the presence of small amounts of unreacted Al and some α-Ti. This observation was further substantiated by examination of the product microstructure in Fig. 7 which shows tiny specks of Al and Ti wetting TiC. No intermetallic titanium aluminides were observed.

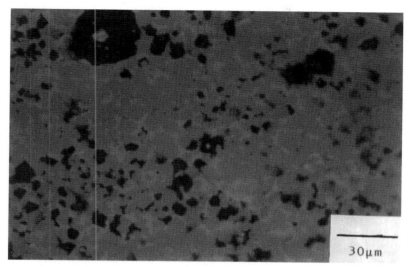

**Fig. 7** *Micrograph from reaction 2 showing unreacted Al and Ti wetting the ceramic particles.*

*Reaction 3*

Ignition of the reaction between Ti and B in the presence of 10 wt% $Al_2O_3$ took place at 1203°C. A plot of temperature against time just after ignition can be found in Fig. 8. The reaction did not self-propagate if the amount of $Al_2O_3$ exceeded this critical amount of 10%. The product microstructure for samples of $TiB_2$-10%$Al_2O_3$ is shown in Fig. 9. The product was extremely porous and consisted of $TiB_2$ particles containing $Al_2O_3$.

## Discussion

*Reaction 1*

It was not surprising that the ignition temperature increased with increasing $Al_2O_3$ dilution because the amount of contact between Ti and carbon black was smaller. As a result, fewer points were available where ignition could occur. As the percentage of $Al_2O_3$ increased, more $Al_2O_3$ was present to absorb heat than Ti and carbon black. This has an important effect on the SHS process which transforms from the thermal explosion mode at 10 wt% $Al_2O_3$ to the unstable SHS mode at 20 wt% $Al_2O_3$. This is evident in Fig. 3 which shows further ignitions while the sample is cooling down after the combustion temperature had been reached. At $Al_2O_3$

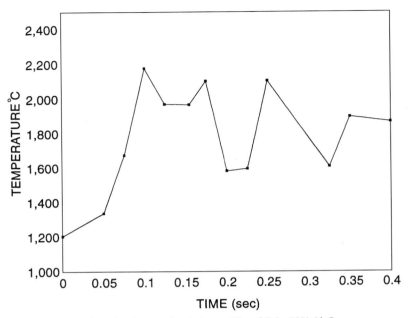

**Fig. 8** *Temperature–time plot for reaction between Ti and B in 10%Al₂O₃.*

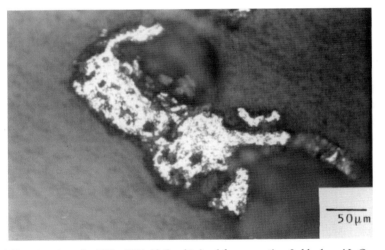

**Fig. 9.** *Microstructure for TiB₂- 10%Al₂O₃ obtained from reaction 3. black = AL₂O₃, white = TiB₂.*

contents in excess of 41 wt%, the reaction was no longer self-propagating. However, by deliberately creating regions of high Ti and carbon black contact within a sample, it was observed that it was still possible for SHS to take place.

It has to be stressed that reaction did take place even if it did not self-propagate. Al$_2$O$_3$ tended to moderate the reaction in such a way as to absorb the exothermic heat such that an additional amount of heat was necessary to make the reaction self-propagating. However, by the time this did happen a thick TiC layer had formed on Ti so that the reaction rate was determined by the rate of solid state diffusion through the product which was slow. The kinetics therefore became so slow that not enough heat could be generated per unit time for the reaction to ignite.

The microstructure of products obtained from SHS had a continuous structure of TiC particles containing Al$_2$O$_3$. This type of microstructure is very interesting because the interfacial bond obtained may be useful in increasing the material toughness.

At a precompaction pressure of 1.36 MPa, the ignition temperature for TiC-20 wt% Al$_2$O$_3$ dropped to 1170°C from 1192°C for loose reactant powders. No ignition took place at precompaction pressures exceeding 2.8 MPa. This observation requires further investigation. It is possible that as the reactant sample was more dense it was likely to be more thermally conducting and therefore conduct heat away from local hot spots to effectively prevent ignition.

*Reaction 2*

This reaction took place in two distinctive steps. The first was the reduction of TiO$_2$ by aluminium

$$3TiO_2 + 4Al \rightarrow 3Ti + 2Al_2O_3$$

which started at 1310°C. The heat released by this reaction increased the temperature to about 1700°C whereupon Ti reacted with carbon black. The resulting microstructure was an intimate mixture of Al$_2$O$_3$ and TiC. The presence of Ti and unreacted Al in the final product together with the absence of TiO$_2$ suggested that some other means of TiO$_2$ reduction had taken place. Some reduction of TiO$_2$ by carbon black occurred at lower temperatures as verified by XRD analysis. This was the endothermic reaction observed by Requena *et al*[14] at 1000°C. Where present, Al and Ti wetted the Al$_2$O$_3$ and TiC particles completely as can be observed in Fig. 7. This effect was not apparent in previous reported studies.[12,14] The presence of unreacted Al and Ti may be at least partly responsible for the improved fracture toughness reported by Adachi *et al*.[13]

*Reaction 3*

It is evident from Fig. 8 that a number of temperature rises took place after ignition. This suggested that the reaction did not take place in one burst but in a series of explosions. This was due to the effect of the Al$_2$O$_3$ dilution, changing the reaction from the thermal explosion mode to one of unstable SHS as explained by Munir.[2] It was therefore not surprising to observe that at additions of only 20 wt% Al$_2$O$_3$ the reaction was not self-sustaining. The microstructure of TiB$_2$–10 wt% Al$_2$O$_3$ consisted of a mixture of continuous grains as well as small individual particles. This was in

131

contrast to pure $TiB_2$ which had formed as continuous rivers of particles. This sharp difference in the microstructure was related to the dilution effect of $Al_2O_3$.

## Conclusions

It was shown that it was possible to produce ceramic composite powders by the three reaction routes which were studied. The rate of reaction of Ti with carbon black and with boron as well as the ignition and combustion temperatures were dependent on the reactant grain size and the amount of $Al_2O_3$. The nature of SHS changed from the thermal explosion mode to unstable SHS as more $Al_2O_3$ was added. The reaction between Ti and carbon black did not follow a self-propagating mode if more than 41 wt% $Al_2O_3$ was present unless the reactants were inhomogeneously mixed. By deliberately introducing regions of high Ti and carbon black concentrations it was possible to synthesise powders containing $Al_2O_3$ in excess of 50 wt%. Precompaction initially reduced the ignition temperature for the reaction between Ti and carbon black by bringing the reactants in closer contact. However, at pressures above 2.8 MPa the reaction was not self-propagating but occurred via solid-state diffusion of carbon to Ti. Reaction 2 produced a different microstructure than reaction 1. While in reaction 1 fine $Al_2O_3$ particles were obtained, reaction 2 yielded $Al_2O_3$ particles which were much larger.

## References

1.  Z. A. Munir: 'Synthesis of High Temperature Materials by Self-Propagating Combustion Methods,' *Ceram. Bull.,* 1988, **67** (2), 342–349.
2.  Z. Munir: 'Reaction Synthesis Processes: Mechanisms and Characteristics,' *Met. Trans. A,* 1992, **23**A, 7–13.
3.  R. A. W. Hill, Le Sutton, R. B. Temple and A. White: *Research,* 1950, **3**, 569.
4.  F. Booth: 'Theory of Self-Propagating Exothermic Reactions in Solid Systems,' *Trans. Farad. Soc.,* 1953, **49**, 272–281.
5.  A. G. Merzhanov and I. P. Boroviskaya: 'Self-Propagating High-Temperature Synthesis of Refractory Inorganic Compounds,' *Dokl. Akad. Nauk. SSSR (Chem),* 1972, **204**, 429–432.
6.  A. G. Merzhanov, A. K. Filonenko and I. P. Boroviskaya: 'New Phenomena in Combustion of Condensed Systems,' *Dokl. Akad. Nauk. SSSR (Phys. Chem.),* 1973, **208**, 122–125.
7.  B. V. Novozhilov: 'Combustion Rate for a Model Two-Component Powder Mixture,' *Dokl. Akad. Nauk. SSSR (Phys. Chem.),* 1960, **144** 1400–1403.
8.  J. B. Holt, D. D. Kingman and G. M. Bianchini: 'Kinetics of the Combustion Synthesis of $TiB_2$,' *Mater Sci. Eng.,* 1985, **71**, 321–327.
9.  A. G. Merzhanov, G. G. Karynk, I. P. Borovinskaya, S. Y. Sharivker, E. I. Moshovskii, V. K. Prokudina and E. G. Dyadko: 'Titanium Carbide Produced by Self-Propagating High-Temperature Synthesis: Valuable Abrasive Material', *Sov. Powd. Met. Ceram.,* 1981, **20** (10), 709–713.

10. O. R. Bergmann and J. Barrington: 'Effect of Explosive Shock waves on Ceramic Powders,' *J. Am. Ceram. Soc.*, 1966, **49**(9), 502–507.
11. M. Ouabdesselam and Z. A. Munir: 'The Sintering of Combustion Synthesised Titanium Diboride,' *J. Mater. Sci.*, 1987, **22**, 1799–1807.
12. M. Abramovici: 'Composite Ceramics in Powder or Sintered Form obtained by Aluminothermic Reactions', *Mater. Sci. Eng.*, 1985, **71**, 313–319.
13. S. Adachi, T. Wada and T. Mihara: 'High-Pressure Self-Combustion Sintering of Alumina-Titanium Carbide Ceramic Composites,' *J. Am. Ceram. Soc.*, 1990, **73**, 1451–1452.
14. J. Requena, J. S. Moya and P. Pena: 'TiC/Al$_2$O$_3$ Powders Obtained by Combustion Synthesis', *Third Euro-Ceramics*, vol. 1, P. Duran and J. F. Fernandez (Eds), Faenza Editrice Iberica S L, 1993, Spain.

# Investigations in Dispersed Phase Systems

S. Zador,[1] S. Dasgupta,[1,2] J. Butler[2] and C.B. Alcock[2,3]

1. The Electrofuel Mfg. Co. Ltd., Toronto, Ontario, Canada
2. University of Toronto, Toronto, Ontario, Canada
3. University of Notre Dame, South Bend, Indiana, USA.

## Abstract

Behaviour of materials in dispersed phase systems has been a major interest of Professor Alcock's over the last forty years. Understanding of the behaviour of materials leads to the development of interesting devices. Two areas of investigations in dispersed phase sytems are reported here. The first area is the study of Ostwald ripening in a molten salt medium. The second area is the development of a device based on a ceramic/ceramic composite which acts as a glow plug for diesel engines.

Ostwald ripening experiments of a dispersed phase in a solid matrix usually yields results suffering from poor statistics and doubts about the dispersoid–matrix contact. Recently, experiments were conducted both on ground and in microgravity to study particle growth in a solid/liquid (FeS particles in a LiCl–KCl) and liquid/liquid (Sb2S3 droplets in LiCl–KCl) dispersions.

The microgravity experiments were carried out in the CSAR–I rocket which allowed 300 s of anneal under zero gravity. A preliminary conclusion from these experiments is that the more rapid ripening in the solid–liquid system when compared with the liquid–liquid system is due to a large difference in the interfacial energies. The $Sb_2S_3$ dispersion appears to be more stable because of this factor alone. Further experiments permitting significantly longer times under micro gravity conditions are now being planned.

The development of an all ceramic glow plug for diesel engines was undertaken to improve the combustion efficiencies of diesel engines and to enhance cold starting. A conducting ceramic dispersed in a solid ceramic matrix was developed which had the appropriate electrical, thermal, chemical and mechanical properties.

Laboratory tests indicated excellent properties of these glow plugs. Recent engine tests at Detroit Diesel Corporation show that diesel engines can be started within 6 s at temperatures of minus 25°C and that the ceramic glow plug has long lifetime under diesel combustion conditions. Commercial production of these glow plugs is underway at Electrofuel.

## OSTWALD RIPENING

## Introduction

Behaviour of materials in a dispersed phase system has been a keen interest of Professor Alcock over the last forty years, and much before this field became fashionable. Professor Alcock's interest in this area had all along been eclectic mix

135

of theoretical studies along with attempts to produce interesting devices. This paper outlines a study on Ostwald ripening in a molten salt medium and the development of a practical device (ceramic glow plugs for diesel engines) based on a ceramic dispersion in a ceramic matrix.

The use of a dispersed not-reactive phase of fine near-spherical particles serves to strengthen metallic alloys by impeding the movement of dislocations which give rise to plastic deformation. When the composite must perform in service for long periods of time at elevated temperatures, it is usually observed that the dispersed-phase strengthening decreases with time. This is because the dispersed phase has a finite solubility in the metallic matrix, and since smaller particles in the dispersion have high surface-to-volume free energy ratios, they tend to dissolve. The dissolved matter deposits on the larger particles in the dispersoid. Since the number of dispersed particles is reduced by this process, known as 'Ostwald ripening', the effect of the dispersion in restraining dislocation movement is diminished.

The theory of particles coarsening by Ostwald ripening has been developed for the case where the solution/diffusion mechanism can either be rate-controlled by dissolution kinetics at the interface between dispersoid and matrix, or by diffusion control of the dissolved matter in the matrix.[1,2,3] Two different time dependences, of these alternative rate-controlling steps, have been proposed which show two different power laws for these cases.

For dissolution control:

$$\bullet \quad r^2(t) - r^2(0) = 64 \text{ K } C_o V_m^2 \gamma / 81 \text{ } RT \qquad (1)$$

and for diffusion control

$$r^3(t) - r^3(0) = 8 \text{ } C_o D V_m^2 \gamma / 9 \text{ } RT \qquad (2)$$

where $K$ and $D$ are reaction rate constants, respectively, $C_o$ is the solubility of a flat sample of dispersoid material in the matrix, $V_m$ is the molar volume and $\gamma$ the surface energy of the dispersoid. The radii $r(o)$ and $r(t)$ are the average radii of the dispersoid initially and after an annealing time of $t$ seconds respectively.

The dissolution control process kinetics should be independent of the overall concentration of the dispersed phase but the diffusion controlled process should reflect this concentration since the higher the concentration of dispersoid, the shorter the average diffusion length between the particles, and hence the more rapid the Ostwald ripening. Care must be taken however that misleading results are not obtained as a result of particle agglomeration at higher concentrations of dispersoid. The theoretical analysis also predicts that the shape of the particle size distribution will depend on the ripening mechanism as shown in Figure 1. It can be seen that the diffusion-controlled process should be distinguishable by the presence of a steep leading edge in the ripened size distribution, whereas the reaction controlled process has a broader distribution.

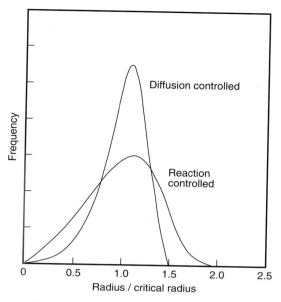

**Figure 1.** *Steady state distribution curves.*

Experiments performed on the practical system of thoria particles dispersed in nickel-chromium alloys at 1350°C showed that the average particle size appeared to conform with the diffusion-controlled process in which the diffusion coefficient of thorium in the matrix was found to be ca $10^{-14}$ cm$^2$ s$^{-1}$ at this temperature.[2] Periods of anneal up to 400 h were required to achieve this result mainly because of this low diffusion coefficient. The particle size distribution will depend on the ripening mechanism as shown in Figure 1. It can be seen that the diffusion–controlled process should be distinguishable by the presence of a steep leading edge in the ripened size distribution, whereas the reaction controlled process has a broader distribution.

The particle size distribution was obtained from electrolytically thinned metal samples using an electron microscope, but because of the difficulties of adequate sampling, the statistics of the experiment (approximately 150 particles were counted after each separate anneal) was not very accurate. Clearly a system such as this would have practical applications because of this very slow Ostwald ripening, but for scientific investigation of the phenomenon, results are hard to obtain which are statistically convincing.

An alternative procedure was therefore devised in which the dispersoid would be dispersed in a molten salt phase. The dispersoid should be a solid of very low solubility in water and molten salt. Here the dispersoid could readily be separated from the matrix by dissolution of the latter in water, and a major obstacle to satisfactory particle statistics could be overcome. It was decided to make a dispersion of solid FeS in a LiCl–KCl eutectic composition matrix at 450°C for initial experiments under normal gravity. Also it was decided to attempt liquid–liquid dispersoid/matrix experiments at 750°C to eliminate any possibility that an effect may arise from a non-spherical average particle shape in the solid–liquid experiments. For these studies $Sb_2S_3$(m. 550°C) was used as the dispersed phase materials; the solubility in this moten salt matrix has been previously determined.

The obvious advantage of using a liquid salt matrix is that the diffusion coefficient of the dissolved species would be expected to be ca $10^{-6}$ $cm^2s^{-1}$ which is several orders higher than the previously discussed solid dispersion in a solid matrix. It is therefore to be expected that the time to achieve significant Ostwald ripening in the liquid matrix would be drastically reduced with the use of liquid matrix. Use of liquid matrix introduces the possible difficulties of interference in the coarsening kinetics by phenomena related to sedimentation of the dispersed phase through the liquid. It was decided to carry out studies under terrestrial conditions to assess these effects under the best possible conditions, and to compare the results with corresponding studies carried out under microgravity conditions, where these disturbing effects should be absent.

## TERRESTRIAL EXPERIMENTS

The first series of experiments were carried put with 0.5 wt% FeS in the eutectic salt matirx. The experimental temperature was 450°C. The change in particle size distribution was observed for ageing times ranging from 1 h to 20 h. The change in mean particle diameter with time was observed and is shown in Table 1. When the cube of the mean particle radius was plotted against time, a linear relationship is obtained (Figure 2). The slope of this curve is found to be $1.16 \times 10^{-16}$ $cm^3$ $s^{-1}$. The

**Table 1: Results of the experiment with 0.5 wt% FeS in the eutectic salt matrix**

| Time (h) | Mean diameter (microns) | Standard deviation |
|----------|------------------------|--------------------|
| 1  | 1.61 | 1.04 |
| 3  | 2.03 | 1.20 |
| 5  | 2.60 | 1.37 |
| 10 | 3.70 | 2.19 |
| 22 | 4.35 | 2.20 |
| 45 | 5.66 | 2.65 |
| 90 | 6.57 | 2.80 |

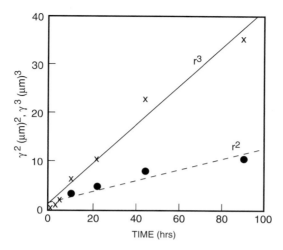

**Figure 2. Salt + 0.5% FeS: $r^3$ and $r^2$ v t plot.**

theoretical slope that can be obtained by substituting values into the equation for diffusion controlled growth can be written as:

$$r(t)^3 - r(0)^3 = [8C_oDV_m{}^2\gamma]/[9RT]$$

where

$$C_o = 4.7 \times 10 \text{ mol cm}^{-3}$$

$$V_m = 18.16 \text{ mol cm}^{-3}$$

$$R = 8.314 \times 10^7 \text{ ergs K}^{-1} \text{ mol}^{-1}$$

$$T = 450°C = 723 \text{ K}$$

$$D = 10^{-5} \text{ cm}^2 \text{ s}^{-1}$$

For which we obtain

$$\gamma = 500 \text{ ergs cm}^{-1}$$

The equilibrium concentration was calculated from data in the literature which as a function of temperature is defined as:

$$\log (X_{FeS}) = -3415/T - 1.112, \qquad (3)$$

Where $X_{FeS}$ is the mole fraction of FeS.[3]

From this, the value of $\Delta r^3/t$ is calculated as $1.07 \times 10^{-17}$ cm$^3$. The difference between the two experimental values of this slope may be attributed to the error in the approximate values of $D$ and $Y$ in the equation. However, when the square of the mean particle radius is plotted against time the straight line seems to show a

139

better fit than that of the cubic graph (Figure 2). This would seem to indicate that at low concentrations it is impossible to conclude from this criterion whether the coarsening mechanism is reaction controlled or diffusion controlled.

The study of the particle size distribution was also inconclusive because the resultant distribution showed too many particles at the large-diameter 'tail' of the distribution for either diffusion or reation control to be predominant (Figure 3).

The second series of experiments were performed with 1.0 wt% ferrous sulphide in the salt matrix which is much higher than the saturation solubility of ferrous

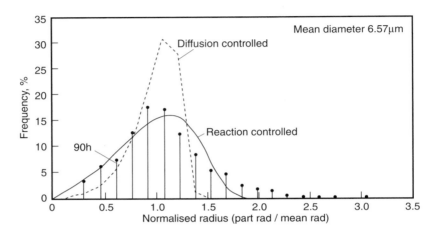

**Figure 3.** *Salt + 0.5% FeS: steady state distribution (90 h).*

sulphide in the salt mixture. The plot of the cumulative frequency of the particles against particle size at different coarsening times is shown in Figure 4). When the square and the cube of the mean particle radius are now plotted against time, the $r^2$ vs. time plot shown a slightly better that the $r^3$ vs. time plot, which would lean

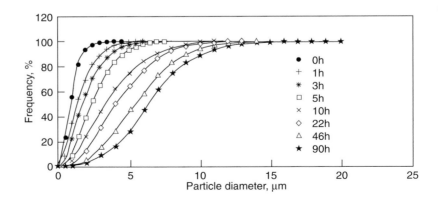

**Figure 4.** *Salt + 1.0 wt% FeS: cumulative particle distribution.*

toward a reaction controlled growth mechanism (Figure 5). This is quite possible, since, as the volume fraction of the particles increase, the mean free path between the particles decreases and hence the diffusion rates might become fast when compared to solution reaction rates which would thus ultimately control the

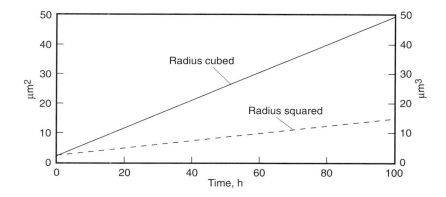

**Figure 5.** *Salt + 1.0 wt% FeS: r3 and r2 v time plot.*

coarsening mechanism. The slope of the curve for $r^3$ is 5.5 x $10^{-16}$, which is higher than that predicted by theory. A plot of the log (*t*) vs. log (*d*), however, gave a slope nearly equal to 3, indicating that diffusion through the matrix is the rate controlling step (Figure 6).

A point to be considered in both of these experiments is that not only is the diffusion mechanism dependent on the cube of the particle radius but the

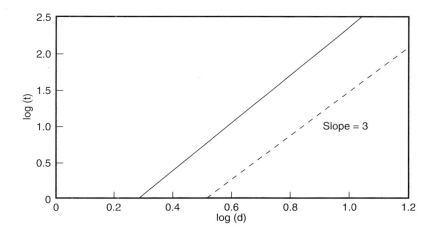

**Figure 6.** *Salt + 1.0 wt% FeS: Log (t) vs log (d) plot.*

141

coalescence of particles with subsequent liquid phase sintering is also dependent on the cube of the particle radius and hence also the indication that collision could be avoided by performing experiments in reduced gravity; therefore a better understanding of the growth phenomenon could be possible.

No terrestrial experiments could be performed with the $Sb_2S_3$ samples as they coalesced into one large droplet in very short time periods.

## EXPERIMENTS IN MICROGRAVITY

Following these ground-based experiments, a parallel set of experiments were prepared for short-term experimentation under microgravity conditions. The equipment to enable as many as 32 experiments to be carried out was constructed using a zinc–silver battery power source of 2400 Wj capacity to heat individually controlled furnaces for each of the 32 samples. The fused salt was contained in electrically heated stainless steel ampoules which could be brought up to operating temperature in 5 s and held at a constant temperature, ± 3°C, for the 5 min duration of the microgravity trajectory (Figure 7). The equipment was launched on the CSAR-1 rocket constructed for the Canadian Space Agency by Bristol Aerospace Ltd., at the NASA facility at White Sands, New Mexico, U.S.A. All experiments were

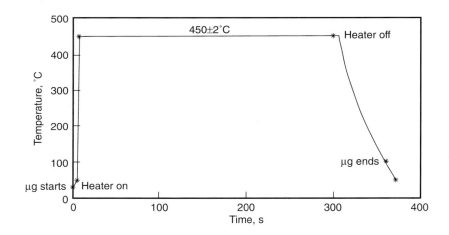

**Figure 7.** *Results from the CSAR-1 experiments for the solid FeS particles dispersed in the eutectic salt matrix.*

satisfactorily conducted and all samples of the dispersoid were recovered from the individual furnaces.

Samples of 0.5,1 and 10% FeS in LiCl–KCl eutectic were annealed at 450°C for a 5-min period of microgravity, and 0.1% $Sb_2S_3$ in the salt matrix were annealed at

750°C for the same period (Figures 8 and 9) show the change in particle size distribution brought about in the FeS samples, and contain comparative results for the terrestrial experiments. The time evolution of the bimodal size distribution follows the general trend to be expected for Ostwald ripening, but the more rapid

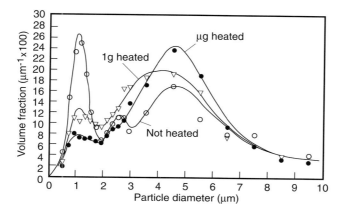

**Figure 8.** *Particle size distribution: 1.0 wt% FeS/LiCl–KCl.*

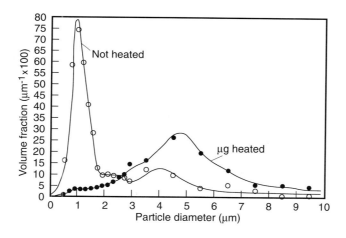

**Figure 9.** *Particle size distribution: 10.0 wt% FeS/LiCl–KCl.*

evolution in the 10% sample demonstrates the effect of increased concentration which is to be expected from a diffusion controlled process. Roughly 20,000 particles were counted in each experiment, thus assuring satisfactory statistics. The results of the $Sb_2S_3$–LiCl–KCl system which was all-liquid at 750°C showed very little ripening during a 5-min anneal (Figure 10), and it was concluded that this reflected a low interfacial energy in the system which would lead to a slow rate of particle

143

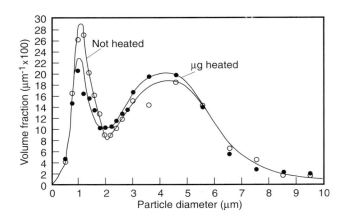

**Figure 10.** *Particle size distribution: 1.0 wt% Sb²S³.LiCl–KCl.*

coarsening. The fact that the original distribution was recovered pratically unchanged from these experiments suggests that there were no extraneous effects which occured during melting or freezing of the samples which might invalidate the results with the FeS experiments.

## CONCLUSION

The results of these experiments strongly suggest that experiments carried out in liquid matrix systems under microgravity conditions can provide valuable information to supersede Earthbound experiments. It is hoped to carry on this series of studies with longer anneals and with a wide range of particle size distributions from a simple Gaussian distribution which might be theoretically tractable, to multimodal distributions which have not been accounted for by theories based on the usual physio-chemical factors such as reaction- or diffusion-control, but a theoretical treatment based on minimum entropy production devised by Kuczynski will be given a closer scrutiny.[6]

A longer duration Ostwald ripening experiment is being planned as a Get Away Special (GAS) experiment on the space shuttle. The flight is planned for December 1995. Over 100 simultaneous experiments will be carried out using a range of variables such as time, particle size distribution, materials and temperature. For good isothermality, the experimental chambers will be built around multiple heat pipes using a sodium potassium melt which is moved around through capillary action. Isothermality better than 0.1°C can be achieved with the heat pipes.

## ALL-CERAMIC GLOW PLUG FOR DIESEL ENGINES

Dispersed phase systems can be used to develop devices with unique and interesting properties. A ceramic dispersion in a ceramic matrix has been utilised to develop a glow plug for use in diesel engines.[7]

Cold diesel engines are difficult to start and combustion assistance is needed to aid compression ignition in indirect Ignition (IDI) diesel engines.[6] Furthermore, the need for glow plugs as an ignition source for alcohol fuelled direct injection(DI) diesel engines and as an aid to cold starting non-conventionally fuelled DI diesel engines has increased in recent years. Recently there is also some speculation that a combustion aid such as the glow plug could improve combustion efficiencies and thus decrease production of pollutants and increase fuel efficiencies.

Traditional glow plugs are pencil shaped, consisting of a tubular heater element (metal/nichrome wire) embedded in magnesia powder inside a metal shealth. A variation of this glow plug uses a silicon nitride sheath. These metal glow plugs suffer from problems related to high corrosion rates, short life, slow response time and low operating temperatures.

Electrofuel developed an all ceramic glow plug which had to meet conditions of high thermal shock, low thermal mass, corrosion resistance within the engine combustion chamber and fast response time within a matter of seconds.

The ceramic composite material based on silicon nitride had a flexural strength $\geq 750$ MPa at 20°C and $\geq 300$ MPa at 1000°C. A fracture toughness $\geq 5.5$ MPa. $^{1}/_{2}$ m at 20°C is capable of withstanding a thermal shock of -50 to 1200°C in 10 s, is able to withstand pressure up to 1500 psi, and is able to withstand mechanical shock loads of more than 40 g.

Typically the glow plug requires about 45 watts to reach temperature. A complete prototype glow plug is presented in Figure 11.

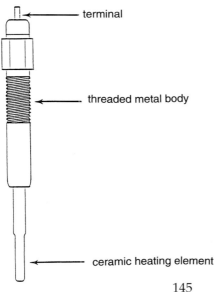

terminal

threaded metal body

ceramic heating element

**Figure 11.**
*A schematic drawing of the all-ceramic glow plug.*

## RESPONSE TIME

The response time for the ceramic glow plug is most important because it controls the cold starting parameters, battery usage and perhaps initial cold engine smoking problems.

Electrofuel's ceramic glow plug is designed to have the fastest response time possible time possible while using the lowest amount of energy from the battery.

Figure 12 shows the response time of the cermaic glow plug.

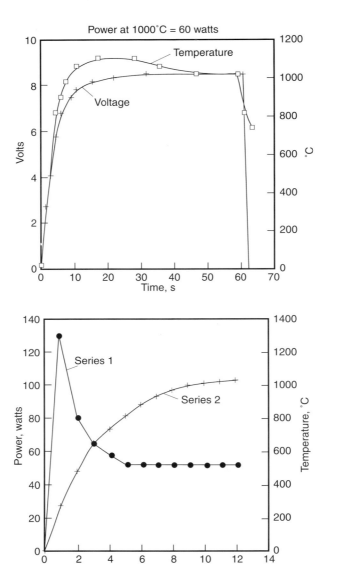

**Figure 13.**
*Performance of the ceramic glow plug showing self-regulating property.*

**Figure 12.**
*Response curve of the ceramic glow plug.*

The Electrofuel ceramic glow plug has a built-in passive control system. Figure 13 shows the power/time graph.

## ENGINE TESTS

Initial iterative and durability tests of the ceramic glow plugs were done at Advance Engine Technology Ltd. (AET) in Nepean, Ontario and at Detroit Diesel Corporation, Detroit (DDC).

The tests at AET used NRC's experimental fuel. This 'worst case' fuel has a high level of aromatic and correspondingly low Cetane number, as well as a high sulphur (0.5%) content. The use of this fuel has been advantageous, since it allows the glow plugs to be subjected to higher amplitude combustion pressure wave vibrations than those experienced with a conventional low aromatic diesel fuel. In addition, the properties of this fuel are well known, providing a basis for comparison with previously published results.

The plugs were installed in the 8V71T engine and subjected to cylinder pressures at idle, to test the plugs for their sealing properties. With a power supply, the plugs were powered to their given specifications.

The next step in the durability tests was to progressively load the engine, subjecting the plugs to increasing loading conditions for extended time periods. The glow plugs were operated continuously. After 500 h of continuous operation at different engine loads and speeds the ceramic plugs showed no sign of wear or corrosion.

With the very encouraging test results at AET, it was decided to send some further glow plugs to Detroit Diesel for further engine testing. Extensive engine tests at DDC (8) further identifies that this ceramic composit glow plug meets the requirements necessary for this critical engine application.[7]

Work is continuing to develop processes through which the ceramic glow plug can be manufactured in a cost effective manner.

## ACKNOWLEDGEMENTS

We would like to acknowledge the support of the following: The Canadian Space Agency, whose focus on basic research under micro-gravity conditions is most encouraging; Transport Canada supported the Glow Plug development work; while the US Department of Energy supported part of the engine testing.

## REFERENCES

1. G.W. Greenwood: *Acta Met.*, 1956, **4**, 243.
2. C. Wagner: *Zeits. Elektrochemie*, 1961, **65**, 581.
3. I. Lifshitz and V.J. Slyozov: *J. Phys. Chem. Solids*, 1962, **19**, 35.
4. P.K. Footner and C.B. Alcock: *Met. Trans.*, 1972, **3**, 2633.
5. R.A. Sharma and R.N. Seefurth: *J. Electrochem. Soc.*, 1984, **131**, 1084.

6.  G.C. Kuczynski: *Sintering Processes*, G.C. Kuczynski, ed. Plenum Press, NY, 1980, p. 3.
7.  S. DasGupta, J.K. Jacobs, S. Radmacher and M. Sobchek 'Ceramic Glow Plug'. US patent 5,304,778, April 19, 1994.
8.  C. Havenith, H. Kuepper and U. Hilger: 'Performance and Emmission Characteristics of the Deutz Glow Plug assisted Heavy Duty Methanol Engine', SAE Technical Paper Series 872245, Truck and Bus Meeting and Exposition, Dearborn, Michigan, November 16–19, 1987.
9.  S. Hinkle: 'Performance and Durability of All Ceramic Glow Plug for Heavy Duty Engines' Anual Automotive Technology Development Contractors', Coordination meeting, US Department of Energy, Vol. 1, M. 661–677, Oct. 1993.

# Hot Corrosion of Coatings for Superalloys

M. G. HOCKING

*Department of Materials, Imperial College, London, UK*

**Abstract**

The composition and hot corrosion of successful coatings alloys for marine gas turbines and for coal gasifiers are discussed.

## Introduction

Hot corrosion occurs in, for example, marine gas turbines and in coal gasifiers, due to the presence of sulphur in the atmosphere.

A barrier film of continuous highly impervious oxide is necessary for protection of alloys from gas corrosion, plus some other method of preventing inward migration of sulphur. A critical concentration of a protective oxide-forming additive like Cr in the alloy will ensure adequate internal oxide precipitation.

Alloys with adequate protection against hot corrosion do not have adequate mechanical properties and so the corrosion-resistant alloys must be used as coatings to protect substrate alloys with good mechanical properties. All aspects of metallic and ceramic coatings have been extensively reviewed elsewhere by the author's group[1].

A major problem in hot corrosion is the formation of molten metal/metal suphide eutectics and coatings alloys are formulated to avoid this. Interdiffusion plays an important role as coalescence of consequent Kirkendall voids reduces coating adhesion and causes coating failure by substrate parting. Al in an aluminide or an MCrAlY coating diffuses inwards into the substrate metal. These and many other such effects are described in reference 1.

## A brief survey of some additive elements in coatings

### Ce

Ce and Y confer resistance to spalling and are often put on as a coating, or ion implanted.[2] Corrosion is more uniform and scale has a higher Cr/Fe ratio than without Ce.[3]

### Al

Sulphidation of Fe–30Nb alloy decreased with increasing [Al], becoming very slow for [Al] > 4.8%.[4] A 35Co35Ni20Cr10Mo alloy also showed decreasing sulphidation with increasing [Al].[5] Internal Al-rich sulphide formed and transport through it governed the sulphidation rate of the alloy.[5] The Fe30Mo sulphidation rate showed a transition to a very low value at critical Al and Mo levels.[6]

In hot $Na_2SO_4$ + NaCl corrosion of Co–Cr–Al alloys, low Al in chromia-forming alloys had little effect; Cr above 20% greatly reduced the corrosion of alumina-forming alloys.[7] NaCl caused scale cracking and increased internal sulphidation.

### Si, Zr

Ni–25Cr–2.5AlX (X = Zr, Hf or Si) alloys in $SO_2$ + $O_2$ at 700°C showed that Si and Zr together were most beneficial in reducing breakaway attack.[8]

### Ta, Ti

Ta or Ti additions to Co–Cr alloys at 900°C in $Na_2SO_4$ + NaCl showed no change in corrosion mechanism and preferential attack down grain boundaries occurred.[9]

### Mo

Mo additions to Co–Cr alloys above 4% caused accelerated attack.[10]

### Nb

Fe-25Cr alloys with Ti and Nb additions show improved scaling resistance to sulphidation but Mn has no effect.[11]

### Miscellaneous

Additions of up to 10% Ti, Hf, Y, V, Nb, Mo or W to CoCrAlY alloy have been ranked in performance during sulphidation.[12] MCrAlX (M = Ni, Fe, Co, Fe+Ni, or Fe+Co) (X = Zr, Hf, Si, or Ta) in $SO_2$ + $O_2$ and in hot sulphate + chloride has been studied.[13] High pressure hot corrosion of NiCr alloys with Ta additions showed enhanced chromia-forming and reduced $Ni_3S_2$ formation.[14] Hot corrosion of MCrAlY has been studied.[15, 16] This survey is not comprehensive due to lack of space.

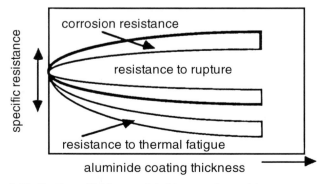

**Fig. 1.** *Aluminide Coatings. Thickness related to corrosion resistance.*

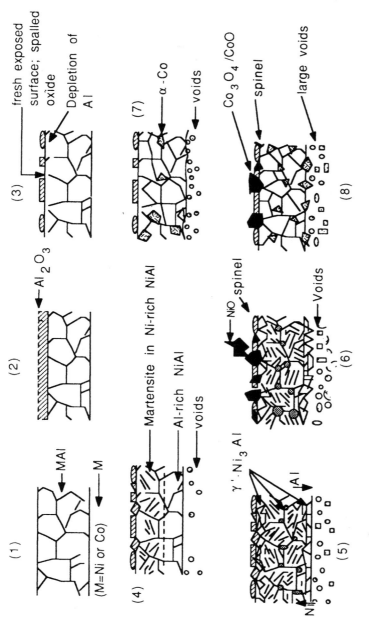

Stages (1)-(3): Common features to both NiAl & CoAl;
Stages (4)-(6): NiAl Coating on Ni Substrate;
Stages (7) &(8): CoAl Coating on Co Substrate

**Fig. 2.** *Morphological Changes in NiAl and CoAl; Alloy Coatings on Ni and Co Substrates respectively.*

151

## Aluminide Coatings

The overall durability of any coating can be assessed as its resistance to corrosion, rupture and thermal cycling. Figure 1 shows schematically the hot corrosion, rupture and thermal fatigue resistance of aluminides as a function of their thickness. Composition changes in NiAl on Ni alloys occur by spalling losses, martensitic transformation and rapid oxidation of the gamma and of gamma-prime phases.[17] Morphological changes in CoAl on Co and in NiAl on Ni are shown schematically in Fig. 2.[18, 19] Failure of aluminide coatings is mainly dependent on the structure and coherence of the alumina layer which is influenced by the available Al activity from the substrate and the transition metal it is alloyed with in the diffusion band.

At first, Al in the aluminide coating diffuses into the substrate but later the diffusion direction reverses and Ni from the substrate starts to enter the coating. This creates kirkendall voids and their coalescence can cause coating disbondment.

Nickel aluminide is still the best aluminide coating for gas turbines. At first, non-protective gamma alumina forms and spalls but then a denser adherent and protective alpha alumina forms. But a damage and repair cycle causes Al depletion in the NiAl below the oxide and may cause precipitation of $Ni_3Al$ which causes NiO to form and to react with alumina to form spinels. The NiO also develops porosity during growth. Appearance of the gamma-prime phase thus marks the onset of a rapid oxidation stage.

Coatings of different morphology are produced by pack or vapour phase methods.[20] In oxidising hot corrosion conditions, the pack and vapour phase aluminides on IN738 had lifetimes of 1500 and 1300 h compared with 300 and 100 h on IN100 and 400 h each on IN939. Substrate influence is evident. In reducing, hot corrosion conditions in sodium sulphide, $NaCrS_2$ and $NaCrO_2$ formed within 1 h, and $NaAlO_2$ appeared after 70 h.

Most high temperature alloys rely on the oxides of Cr and Al for an overall resistance to hot corrosion. Figure 3 shows the solubility of additives and their diffusion barrier tendency, ranging from about 30% solubility for Pt to 1% for Mo. If Al diffusion towards the substrate is arrested, the aluminide coating is an effective barrier for hot corrosion and this occurs when the cumulative content of Cr, W, Ti, Ta, Nb and Mo is over 20%. For most alloys, a high Cr level is needed to compensate for the absence of a low level of these other elements.

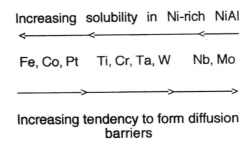

Fig. 3. *Solubilities of elements in Ni-rich NiAl.*

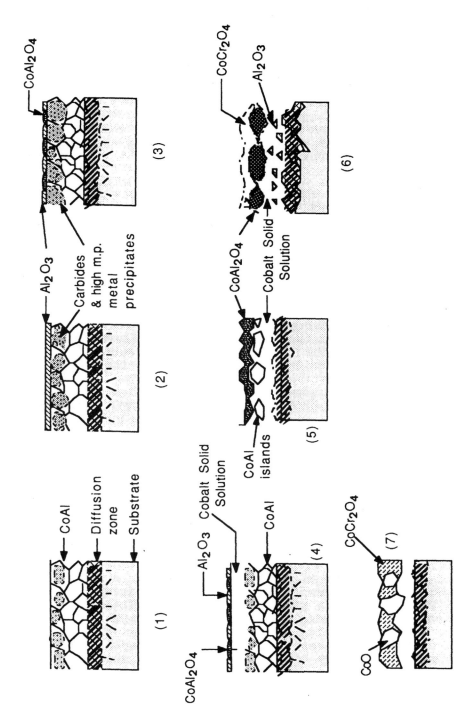

**Fig. 4.** *Morphological changes in a CoCrAlX coating during corrosion/oxidation.*

Ni base alloys form Cr-rich intermediate layers during aluminising, unlike Fe base alloys.[21] Fe–25Cr was the most resistant to sulphide-forming and sulphur-containing environments below 800°C where Ni and Co base alloys fail.[22] The presence of Cr in the aluminide layer arrests spalling and improves the coating microstructure reducing pitting of the coating. It also restricts diffusion of Mo, W and V from the substrate and their consequent oxidation.[23] But an internal sulphidation zone of the spinel $(Cr, Al)_3S_4$ can form, with a wide Cr : Al range.[24]

## Complex Aluminide Coatings

Al coatings are rarely used alone and dual systems like Ni–Al, Co–Al, Cr–Al, or Pt–Al or more complex systems with four to six elements are used to resist hot corrosion, such as MCrAl–X1 or MCrAlY–X2, where M = Ni, Co or Fe; X1 and X2 = Ce, Hf, Pt-group, Si, Ti, Y, Zr etc. Gas turbine coatings are mainly NiCrAl–X1–X2 and CoCrAl–X1–X2 while coal gasifier coatings are mainly FeCrAl–X1–X2 to avoid the nickel/nickel sulphide eutectic which forms especially in reducing sulphurous conditions. Figures 4 and 5 are schematic of the non-sulphidation aspect of the hot corrosion of CoCrAlX and NiCrAlX (sulphides not shown). Figure 6 includes sulphide formation.

Cr is essential in all aluminide coatings, improving the stability, coherence and continuity of the alpha-alumina layer which forms as the top scale or just below a chromia layer.

The hot corrosion of MCrAlX coating alloys with Cr at 20–35% and Al at 2–5% has been investigated[25] where M is Ni, Fe, Co, Fe+Ni or Fe+Co; X is Zr or Hf, with 1.5% Si, 1–4%Ti and 3–4%Ti or 11%Ta. The aim was to find a coating to resist both 'low' and 'high' temperature hot corrosion. 'Low' temperature (around 700°C) hot corrosion occurs, for example, with Co-rich alloys due to the existence of a low melting $CoSO_4+Na_2SO_4$ eutectic, while such alloys perform very well at higher temperature because the pressure of $SO_3$ needed to form the $CoSO_4$ is too great at high temperatures for it to form, so the damaging eutectic cannot exist.[26] A similar mechanism for Fe alloys has been proposed.[27] Specimens coated with $NaCl+Na_2SO_4$ and in $SO_2+O_2$ mixtures showed the bad effects of high Ni at 700°C. Adding 0.5–2%Si and Zr or Hf delayed the onset of accelerated attack. In contrast, FeCoCr–AlX showed low corrosion rates at 700°C and negligible sulphidation. Ti and Si additions conferred good scale adherence. Ta was effective in reducing the scale sulphide content while enhancing chromia formation. At 900°C, hot corrosion was much less for all the alloys due to decreased sulphide stability and formation. Coatings have been developed with adequate performance at both 700 and 900°C.[22, 25, 28, 29, 30, 31]

Iron base alloys can accept up to 5% Al before fabrication problems occur. Fe(15–25) Cr4Al is principally an alumina former,[32] while the Ni and Co base superalloys form mainly chromia.

Yttrium is the only very widely investigated additive in all three systems, FeCrAl, NiCrAl and CoCrAl. Corrosion of MCrAlX systems is mostly the failure of the MCrAl scale. Y has a very beneficial effect on scale retention, especially in thermal cycling.[33, 34]

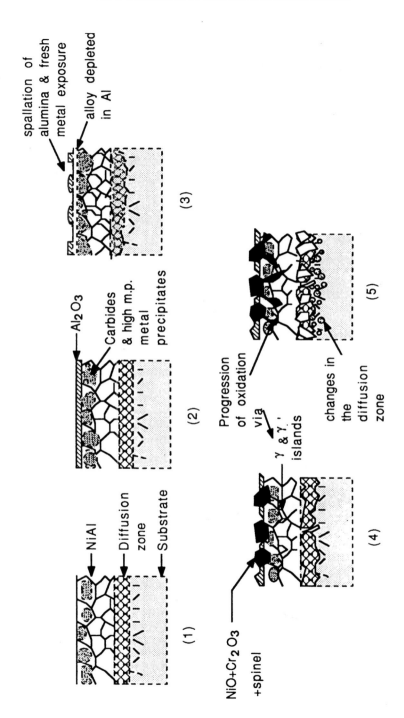

**Fig. 5.** *Morphological changes in a NiCrAlX coating during corrosion/oxidation.*

155

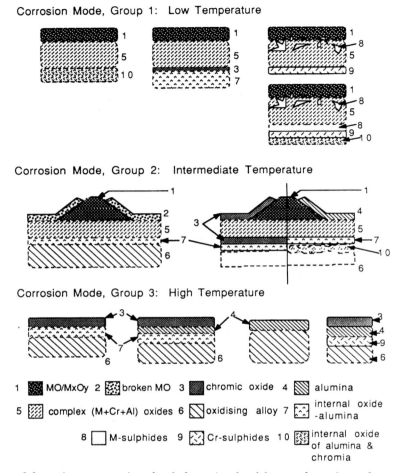

**Fig. 6.** *Schematic representation of scale formation, breakdown, reformation and compositional changes occurring in NiCrAl, FeCrAl and CoCrAl coatings.*

Factors needing to be investigated include how stress generation modes change, how scale plasticity is affected at the metal/scale interface, how vacancy void formation is affected, how preferential growth in an inwards direction into the substrate occurs and how cracks propagate.[34]

Additives like Y, Ce, Hf Zr etc are called active additions and their effects have been summarised from various studies:[35]

(i)   Y improves scale plasticity by changing its microstructure;
(ii)  oxide pegs grow, which anchor scale to substrate;
(iii) mechanical property differences are graded between scale and substrate;

(iv)   scale growth mechanisms are affected by influencing the diffusion parameters, thus modifying the transport and growth of the barrier layer;

(v)   chemical links between scale and substrate are improved.

Y is only beneficial at low levels and is usually added at less than 1%[36] and its sulphur gettering role has been studied.[37] Y and Hf decrease the corrosion rate of NiCrAl, NiCrAlTi and NiCrAlZr alloys, but not over the full 600–800°C range because of liquid product formation.[38, 39] Many permutations of CoCrAlY and NiCrAlY coating systems have been studied and Hf has been a prime additive element.[40, 41] Zr, Si, W, Mo, Mn, Ce etc have also been tested as additives.

Hot corrosion by $Na_2SO_4$ on CoCrAlY at 700°C in $SO_2+O_2$ mixtures has been reported to be due to hole formation in the oxide scale, by a reaction between yttria and $Na_2SO_4$, but this view is not supported by similar hole formation developing in alloys without Y.[42] At 1100°C CoCrAlY coatings showed very good corrosion resistance to $V_2O_5+Na_2SO_4$ and $NaCl+Na_2SO_4$ synthetic corrosion ash[43] while Ni-base alloys are unsuitable, showing catastrophic attack. Creep rupture and cyclic load tests have showed that a high Co content is needed for scale retention on aluminides, chromide aluminides and MCrAlY–X coatings on IN100 substrate.[44] Other additives like Hf, Ce, Zr and W decrease the corrosion rate in general.

## Silicide Coatings

Si is the third choice of additive for its ability to form a protective silica barrier, offering the best melt fluxing resistance in high $pSO_3$ environments. Although normally very stable, degradation of silica occurs by cracking, spalling or by vaporisation as SiO gas.

Si-containing protective scales have been observed on Ni–5Si and Ni–11Si, a chromia-silica scale on Ni40Cr5Si and Ni40Cr11Si and an alumina-silica scale on CoCrAlY–5Si and CoCrAlY–12Si. An Al–12Si alloy forms a eutectic at 577°C and an alumina–silica reaction gives mullite at high temperatures. Si diffuses inwards into the substrate alloy, more so at low $pO_2$. Silica is reduced by outward diffusing Ti and a barrier is needed. Si in NiCrAlY and CoCrAlY systems gives a slight improvement in their oxidation in air with NaCl present but spalling occurs at 700°C and 900°C.[45–50]

Discussions of burner rig data and of thermal barrier coatings are beyond the scope of this short paper but are detailed in reference 1.

## References

1.   M. G. Hocking, V. Vasantasree and P. S. Sidky: 'Metallic and Ceramic Coatings', Longman, UK, 1989.

2.   V. A. C. Haanappel et al, Ox. Met., 1991, **35**, 405.

3.   M. F. Stroosnijder et al, ibid, 1991, **35**, 19.

4.   G. Wang et al, Ox. Met., 1991, **35**, 279.

5.  Chichang Shing *et al*, ibid, 1991, **35**, 295.
6.  G. Wang *et al*, ibid, 1991, **35**, 349.
7.  V. Nagarajan *et al*, *Corros. Sci.*, 1982, **22**, 429.
8.  K. N. Strafford *et al*, *Corros. Sci.*, 1989, **29**, 673.
9.  V. Nagarajan *et al*, *Corros. Sci.*, 1982, **22**, 407.
10. idem, ibid., 1982, **22**, 441.
11. C. R. Wang *et al*, *Ox. Met.*, 1990, **33**, 55.
12. K. N. Strafford *et al*, 1989, **29**, 775.
13. K. N. Strafford *et al*, *Proc. US/UK Wkshp. on Gas Turbines*, Bath 1984.
14. U. Ma *et al*, *Proc. 9th Intl. Cong. Met. Corros.*, Toronto, 64, 1984/2.
15. M. G. Hocking *et al*, *Proc. Conf. on Molen Salts/Metals*, York 1986.
16. C. B. Alcock, M. G. Hocking and S. Zador, *Corros. Sci.*, 1969, **9**, 111.
17. J. L. Smialek and C. E. Lowell, *J. Electrochem Soc.*, 1974, **111**, 800.
18. M. G. Fontana and R. W. Staehle, *'Advances in Corrosion Science and Technology'*, Vol 6, Plenum, NY, 1976.
19. A. R. Nicoll, *'Coatings for High Temp. Oxidation and Hot Corrosion'*, CEI Course on High Temp. Materials and Coatings, June 1984 Finland.
20. P. A. Mari, J. M. Chaix and J. P. Larpin, *Oxid. Met.*, 1982, **17**, 315.
21. E. Fitzer and H. J. Maurer, *'Materials and Coatings to Resist High Temperature Corrosion'*, D. R. Holmes and A. Rahmel (Ed), *Appl. Sci.*, 1977, 253.
22. K. Upadhya and K. N. Strafford, *'High Temp. Protective Coatings'*, *Proc. Symp. Atlanta, GA*, S. C. Singhal (Ed), *Metall. Soc. AIME*, 1983, 159.
23. E. Godlewski and E. Godlewska, *Mat. Sci. Eng.*, 1987, **88**, 103.
24. E. Erdos and A. Rahmel, *Oxid. Met.*, 1986, **26** (1/2), 101.
25. K. N. Strafford and W. Y. Chan, *Proc. 8th Intl. Conf. on Met. Corros.*, Mainz, 1981, 1476.
26. K. L. Luthra, *Conf. on High Temp. Corrosion.*, 1983, Vol NACE-6, NACE, Houston, 507.
27. V. Buscaglia, P. Nanni and C. Bottino, *Corros. Sci.*, 1990, **30**, 327.
28. J. F. G. Conde and G. C. Booth, *Conf. on Deposition and Corrosion in Gas Turbines*, CEGB, Sunbury House, London, 1972.
29. M. G. Hocking and V. Vasantasree, *'Studies on some Basic Mechanisms of Hot Corrosion,' Proc. 4th US/UK Conf. on Gas Turbine Materials in a Marine Environment*, Annapolis, USA, 1979.
30. M. G. Hocking, V. Vasantasree and A. H. Wai, *Proc. 31st Mtg. Intl. Soc. Electrochem*, Venice, 1980.
31. J. F. G. Conde, N. Birks, M. G. Hocking and V. Vasantasree, *Proc. 4th US–UK Conf. on Gas Turbine Materials in Marine Environment*, Annapolis, USA, 1979.
32. M. J. Bennett, M. R. Houlton and G. Dearnaley, *Corro. Sci.*, 1980, **20**, 69.
33. P. Lacombe, *Mat. Sci. Eng.*, 1987, **96**, 1.
34. J. Stringer, A. Rahmel, G. C. Wood, P. Kofstad and D. L. Douglass, *Oxid. Met.*, 1985, **23**, 251.
35. P. Lacombe, *Mat. Sci. Eng.*, 1987, **96**, 1.
36. P. Choquet, C. Indrigo and R. Mevrel, *Mat. Sci. Eng.*, 1978, **88**, 97.
37. J. G. Smeggil, *Mat. Sci. Eng.*, 1987, **87**, 261.

38. J. G. Smeggil and N. S. Bornstein, R78-914387-1, United Tech. Res., Center Rep. (1978).
39. R. L. Jones and S. T. Gadomski, *J. Electrochem Soc.*, 1977, **124**, 1641.
40. M. G. Hocking and V. Vasantasree, 'High Temp. Corrosion of Ni-Cr-Al-Y Alloys in Molten Sodium Sulphate + Chloride Mixtures,' *Proc. Conf. on Interaction of Molten Salts and Metals*, York, 1986, pp 284-291.
41. M. A. M. Swidzinski, PhD Thesis, Univ. of London (1980).
42. S. Y. Hwang *et al*, 'High Temp. Protective Coatings,' *Proc. Symp. Atlanta, GA*, S. C. Singhal (Ed), *Metall. Soc. AIME*, 1983, 121.
43. M. Nakamori, Y. Harada and I. Hukue, 'High Temp. Protective Coatings,' *Proc. Symp. Atlanta GA*, S. C. Singhal (Ed), *Metall. Soc. AIME*, 1983, 175.
44. K. Schmitt, H. G. Thomas and G. Johner, *Proc. 8th Internat. Congress on Metallic Corrosion*, Mainz, Sept 1981, 724.
45. P. S. Sidky and M. G. Hocking, 'Corrosion of Ni-Base Alloys in $SO_2+O_2$ Atmospheres at 700°C and 900°C,' *Proc. Conf. on Behaviour of High Temperature Alloys in Aggressive Environments*, Petten, Netherlands (1979).
46. P. S. Sidky and M. G. Hocking, 'Hot Corrosion of Ni Based Ternary Alloys and Superalloys for Gas Turbine Applications I: Corrosion in $SO_2+O_2$ Atmospheres,' *Corrosion Sci.*, 1987, **27**, 183.
47. P. S. Sidky and M. G. Hocking, 'Hot Corrosion of Ni-based Ternary Alloys and Superalloys for Applications in Gas Turbines Employing Residual Fuels,' *Corrosion Sci.*, 1987, **27**, 499.
48. P. S. Sidky and M. G. Hocking, 'Hot Corrosion of Ni Based Ternary Alloys and Superalloys for Gas Turbine Applications I Corrosion in $SO_2+O_2$ Atmospheres', *Corrosion Sci.*, 1987, **27**, 183.
49. J. B. Marriott, M. Merz, P. R. Sahn and D. P. Whittle (Eds), *The Metals Soc.* UK, 1980, 243, 813.
50. V. Vasantasree and M. G. Hocking, Internal Report, Unpublished work, 1984.

# Thermodynamic calculations for the growth of alumina by directed melt oxidation

F. J. A. H. Guillard, R. J. Hand, W. E. Lee and B. B. Argent

*Department of Engineering Materials, University of Sheffield S1 4DU, UK*

## Abstract

Equilibrium thermodynamic calculations using the computation package MTDATA on the Al/Mg/Si/N/O system are used to predict potential compound formation during the production of $Al_2O_3$-Al composites. These predictions are shown to be in broad agreement with what happens during the production of alumina from aluminium via the DIMOX™ process; in particular it is apparent that intermediate nitrides and a silicide are formed during this process.

## Introduction

The DIMOX™ (DIrected Metal OXidation) process, developed by the Lanxide Corporation, Delaware, USA, involves the continuous infiltration of a predetermined region, which may be empty or contain a porous mass of reinforcement or filler (preform), by the outward oxidation of suitably doped molten metals[1-3]. The dopant elements are of fundamental importance in determining whether growth occurs by directed melt oxidation or not; in the $Al/Al_2O_3$ system the presence of Mg is believed to control the basic oxidation process [4-6] whereas Si and other group IVB elements (Ge, Sn or Pb; C is excluded due to its tendency to form carbides[7]) are believed to accelerate growth. How these dopants affect growth depends on the chemical reactions that occur between them, the parent metal and atmospheric gases. The potential equilibrium reactions between Al, Mg and Si in the presence of O and N have been studied using the MTDATA package[8] and are presented below; the results of these calculations are compared with experimentally observed reaction sequences in the $Al/Al_2O_3$ system doped with Mg and Si.

## Experimental

*Methods*

Ultra-pure (99.9999 wt%) aluminium (Alcan) with Si, Mg and MgO as dopants was used to produce alumina matrix composites via the DIMOX™ process. 0.05–0.06 kg blocks of Al (30 x 25 x 25 mm) had their top surfaces ground on emery paper and were cleaned in acetone to remove any surface contamination. They were then placed in moulds shaped in alumina crucibles using hydroxyapatite powder (Fig. 1). The hydroxyapatite acts as a barrier and directs the reaction upwards and contains it within the space above the aluminium ingots. A layer of dopants (Fig. 1a) comprised

161

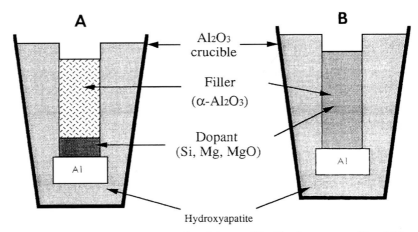

**Fig. 1** *Arrangement of constituents in the $Al_2O_3$ crucible. The dopants are either (a) layered on the metal surface or (b) mixed with the alumina filler.*

of 3 wt% of each of reagent grade Mg and MgO (Fisons) and ultra-pure Si (Wm Rowland), ground using a mechanical pestle and mortar, was placed on top of the Al block. The region above this was filled with a powdered alumina preform; powder X-ray diffraction revealed the presence of about 5–10 vol. % Na $\beta$-$Al_2O_3$ in this preform. An alternative approach mixed the dopants each at 5 wt% concentration level relative to the aluminium with the alumina preform (Fig. 1b).

The oxidising process was carried out in air in an electric furnace. Samples were ramped at 200°C/h up to a soaking temperature of 1453 K which was maintained for 5 h, 15 h or 50 h. To examine the early stages of microstructural evolution, oxidation was sometimes carried out at 1223 K or 1073 K with such samples being cooled directly from temperature. In all cases samples were left to cool to room temperature inside the furnace.

Quantitative phase analysis by X-ray diffraction (XRD) was performed on bulk samples sectioned in the growth direction. To preserve the integrity of the growth product, some samples were embedded in Araldite prior to X-ray diffraction. Other cross-sections were observed under polarised light using a Polyvar optical microscope. Identification and distribution of phases in the samples were more precisely investigated using a Camscan Series 2A scanning electron microscope (SEM). Microanalysis was performed with Link AN10000 or Link EXL EDS (energy dispersive spectroscopy) systems.

*Results*

*Layered Dopants*
Growth occurred without an incubation period; rapid infiltration of the dopant layer occurred with the resulting oxidation products after 5 h at 1453 K being inhomogeneous due to the layering of the dopants on top of the aluminium block. The growth product essentially consists of two layers, the position of the lower of which corresponds to the site of the original layer of dopants. Transmission electron

162

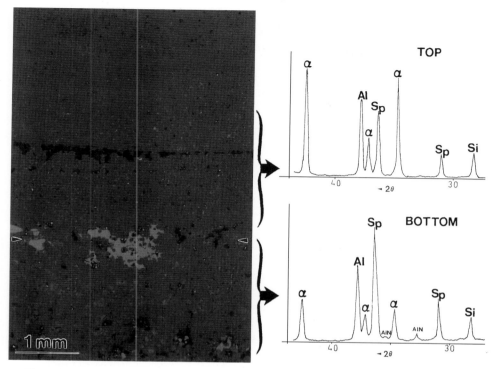

**Fig. 2** *SEM image and XRD patterns from the top and bottom regions of a layered sample after 5 h at 1453 K: $\alpha = \alpha - Al_2O_3$, $Sp = MgAl_2O_4$ spinel, $Si = $ silicon, $AlN = $ aluminium nitride.*

microscopy (TEM) studies confirmed that there was no filler in this region. This lower layer is very heterogeneous but is comprised mainly of Al and spinel with some alumina and traces of AlN (Fig. 2). Meanwhile the top layer is predominantly alumina and Al. There are also traces of spinel but, as these are only detected by XRD, they may arise from overlap with the bottom layer during XRD. Si seems to be evenly distributed in the sample.

The bottom layer can be further divided into two regions: near the remaining Al-ingot, it consists of small areas of spheroids dispersed in Al. The spheroids are believed to arise at the site of agglomerates of dopant particles. The region above this contains large, but variable sized, aluminium inclusions containing Si precipitates characteristic of a hypo-eutectic Al–Si alloy in addition to a matrix of alumina and spinel containing multiple metal channels. The growth product in this region thus has a complex structure varying over the sample. The spheroids have predominantly spinel cores as shown by the X-ray dot map performed on a spheroid located in a spinel-rich region (Fig. 3). Aluminium metal and silicon are also present in the cores but are mainly concentrated on the rim of the spheroids (Figs 3b and 3c). Moreover, the level of porosity is higher inside the spheroids than in the surrounding matrix. In some cases the metal is hardly present in the core, suggesting that the metal present inside the spheroid has been mostly consumed

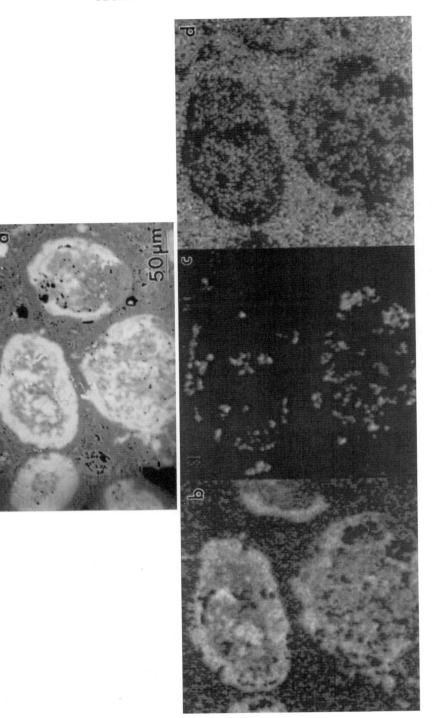

**Fig. 3** (a) SEM image and X-ray dot maps from the bottom layer in a layered sample after 5 h at 1453 K; (b) Al, (c) Si, (d) Mg.

by formation of spinel. Small concentrations of nitrogen were also detected in some spheroids by micro-analysis; this is associated with the presence of aluminium nitride identified by X-ray diffraction. Although nitrogen was only detected in the spheroids the difficulty of nitrogen detection means that it cannot be firmly stated that it is only present within them although this seems quite likely.

Microstructural analysis shows that the top layer (Fig. 4) is composed of Si randomly intermixed in $Al_2O_3$ or Al with no spinel being detected in the SEM by EDS. This layer is porous, containing two types of pore: large pores due to hydroxyapatite contamination and small pores with the same shape as the metallic inclusions which are assumed to result from their oxidation.

After 50 h at the same temperature (Fig. 4), the Al-ingot is totally exhausted and the top layer has thickened. The lower layer has not changed in thickness but the Al inclusions have disappeared so that it now consists of Si-containing $Al_2O_3$ and spinel irrigated by Al-Si channels. The spheroids are unchanged.

Heating instead to the lower temperature of 1073 K led to the formation of MgO, spinel and $Mg_2Si$. Intermediate temperatures showed the gradual elimination of $Mg_2Si$ and MgO.

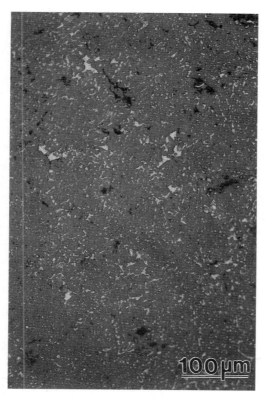

**Fig. 4** *Optical micrograph of the top layer of a layered sample after 50 h at 1453 K which is similar to the microstructure from the mixed dopant layer after 15 h at 1453 K. The bright phase is Si, the dark regions pores and the grey regions $Al_2O_3$.*

*Dopants mixed with filler*

An incubation period was observed and no growth was observed for samples oxidised for less than 15 h at 1453 K although XRD showed that the magnesium had been entirely converted to $MgAl_2O_4$ by reaction with the alumina powder. After 50 h the aluminium was entirely exhausted and XRD indicates that $Al_2O_3$ and Al are the dominant phases while spinel and silicon are only present in small amounts. The structure is more homogeneous than that found with the layered dopants and consists of a porous alumina matrix with spinel and silicon dispersed in it and which contains Al/Si alloy channels (much like Fig. 4).

*Additional Results*

Although the majority of the experiments performed in this study involved growth into preform bodies some additional experiments without a preform were also performed, i.e. simply placing a layer of dopants on the aluminium block without a filler present. In this case after 5 h at 1453 K in a layered sample growth had occurred but only one layer had developed. The microstructure of this layer is identical to that found in the lower layer of the layered dopant samples with preform. Therefore the preform body does not affect the basic growth patterns (most other published studies have looked at growth into systems without preforms); instead it is the dopants that are the crucial part of the process as without them no growth occurred after 24 h at 1453 K.

## Thermodynamic Modelling

As the DIMOX™ process is not an equilibrium process considerable caution has to be exercised in modelling it with equilibrium thermodynamics. The equilibrium structures can be predicted and when an expected phase is not observed, its absence points to the existence of kinetic constraints. Information on the thermodynamics of the Al–Mg–Si system was extracted from the SGTE (Scientific Group Thermodata Europe) solution data base SGSOL and on other possible substances from the substance database SGSUB.[9] All calculations were made using the MULTIPHASE module of MTDATA.[10]

In all cases the systems examined contained aluminium, silicon, magnesium, oxygen and nitrogen. All calculations were carried out assuming atmospheric pressure and a fixed temperature.

It was decided to model the process for the two conditions used experimentally, i.e.

(a) where the dopants are mixed and placed in a layer on top of the aluminium, underneath the alumina filler (Fig. 1a);
(b) where the dopants are mixed with the filler and placed on top of the aluminium (Fig. 1b).

In case (a) it is assumed that solution of magnesium and silicon into the aluminium takes place rapidly and that the process can be considered as the reaction of the aluminium plus dopants with an increasing quantity of the alumina filler under an oxygen gradient starting at a very low level near the start of the filler and

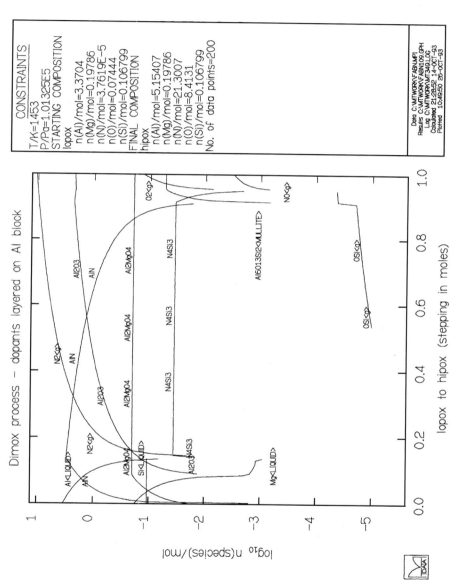

**Fig. 5** *Predicted phase constitution for layered samples as the overall composition changes from that typical of the area close to the original aluminium block, under conditions applying early in the process, to the total system when fully oxidised; situation at 1453 K.*

167

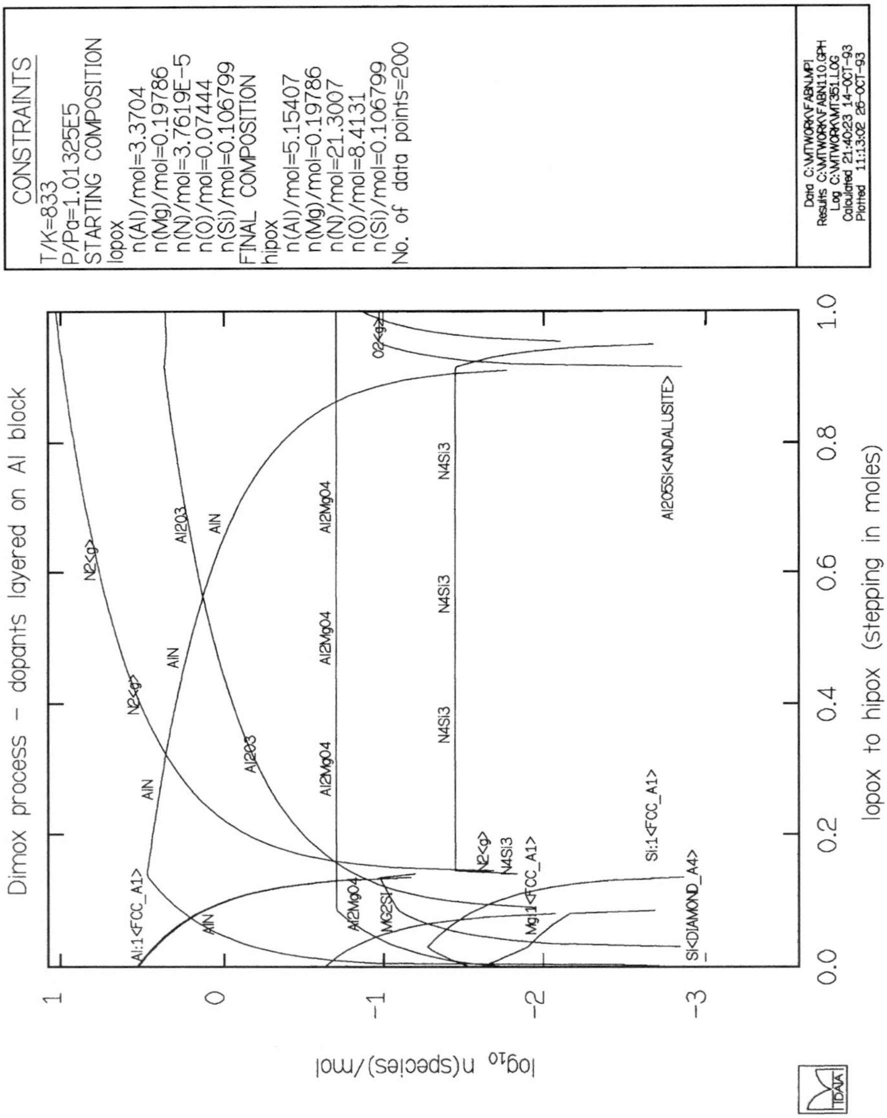

**Fig. 6** *Similar to Fig. 5 but after cooling to 833 K.*

increasing to 5% more than required to oxidise the metallic components at the free surface. Figure 5 shows the predictions at 1453 K and Fig. 6 shows how the phase constitution will change on cooling to 833K where it might be reasonable to consider that the equilibria are 'frozen in'.

It must be emphasised that the diagrams describe the system as the overall composition changes from that typical of the area close to the original aluminium block, under conditions applying early in the process, to the total system when fully oxidised.

The oxygen gradient is reasonable as there is undoubtedly an excess of oxygen at the free surface and under conditions of limited oxygen supply to the metal surface and efficient gettering by the metallic components the starting oxygen potential should be low.

Complex structures are to be expected near to the original interface between the aluminium and the dopants and extending some way into the alumina filler, followed by an extended region in which no further formation of spinel is expected and where silicon-containing phases will only be formed if the reaction rates are insufficient for formation to occur fully at the time when sufficient nitrogen is available to form the nitride.

In case (b) the starting position is pure aluminium which then reacts with increasing amounts of dopants plus filler with an oxygen gradient similar to that described for case (a). Figure 7 shows the predicted situation at 1453 K and Fig. 8 shows the change on cooling to 833 K. The diagrams suggest the development of a more homogeneous structure with spinel formed throughout the sample. Regions which were penetrated by molten aluminium are expected to show silicon precipitates but not $Mg_2Si$ as predicted for the layered dopants.

Equilibrium is clearly not obtained under the experimental conditions used as transport of oxygen into the centre of the samples will be slow. The microstructures observed experimentally in the bulk of the samples are likely to match the predictions for intermediate oxygen levels.

## Discussion

The thermodynamic calculations presented above only indicate which compounds are likely to be formed within the DIMOX™ process. They do not deal with any kinetic effects and in particular ignore phenomena such as wetting behaviour which will influence bulk motion during growth and the resultant distribution of reactants and products at the end of the reaction sequence.

*Layered dopants*

In addition, in the layered dopant system considered here, the reactions within the dopants layer and the complexity of the microstructure within this layer will be further modified by any local compositional variations within it. The transformations in the heterogeneous lower layer of the layered dopant system that occur with time are ascribed to the oxygen trapped in this layer as it is thought unlikely that O will diffuse through this layer as it will preferentially react with the

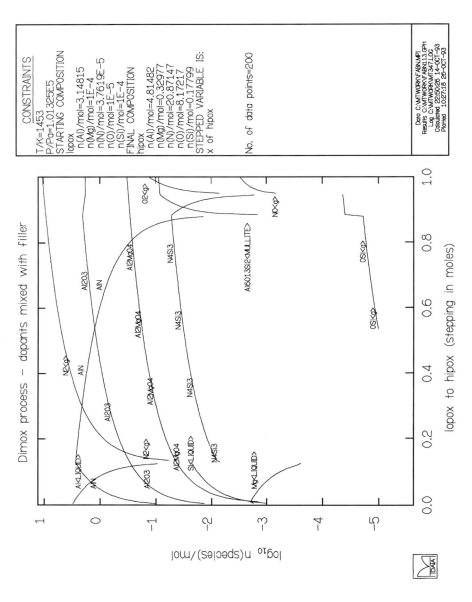

**Fig. 7** *Predicted phase constitution for samples where the dopants were mixed with the filler as the overall composition changes from that typical of the area close to the original aluminium block, under conditions applying early in the process, to the total system when fully oxidised; situation at 1453 K.*

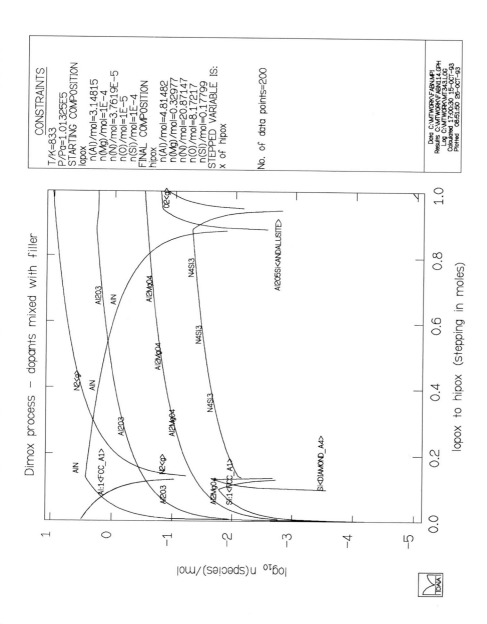

**Fig. 8** *Similar to Fig. 7 but after cooling to 833 K.*

Si within the layer. Thus there is a limited oxygen supply within this layer and this corresponds to the thermodynamic calculations at the 'lopox' (low oxygen) end of the diagram where insufficient oxygen was included to completely oxidise all the elements present.

Depending on the concentration of Mg, alumina or spinel are formed leaving residual metal inclusions where there is incomplete oxidation. Al channels, mainly within the alumina and often containing dissolved Si, develop both from these patches and the reservoir. The spheroids are believed to be MgO particles which, on contact with Al, form intermixed spinel at the metal/MgO interface with further reactions resulting in porosity due to the oxidation of the remaining metal in this region. Once the oxygen in this region is totally consumed, Al reacts with the nitrogen also trapped on infiltration and forms AlN. This is in line with the thermodynamic calculations which show that $O_2$ is primarily used to form $MgAl_2O_4$ in the system Mg, Al, O, N with the rest of the oxygen reacting with Al to form alumina unless the atmosphere is deficient in oxygen. In the latter case AlN becomes the stable phase in equilibrium with $MgAl_2O_4$, $Mg_2Si$, free silicon, and aluminium containing magnesium and silicon in solution.

With time, Al–Si channels wick through the reacted dopant layer and reach the filler powder where actual composite growth proceeds. The bottom layer continues to evolve with the large metallic zones within it being converted to spinel or alumina. The first stage can then be regarded as an initiation period and it is only after this when the dopant layer has been completely penetrated by the molten metal that true composite growth occurs in the filler.

As more air becomes available the silicon is predicted initially to be fixed as silicon nitride although at a concentration that is probably below the XRD detection limit. It is noticeable that both the silicon and the magnesium are predicted to be exhausted within the bottom 20% of the composite. In practice spinel is not observed in the top part of the sample but some silicon has been shown to be present by EDS, suggesting that the formation of silicon nitride from silicon in solution may be slow.

*Dopants mixed with filler*
The homogeneous structures containing alumina and spinel have been predicted but the slow approach to equilibrium leaves channels containing aluminium. Silicon is predicted to form silicon nitride or silica depending on the oxygen potential but it appears that reaction rates are sufficiently slow to leave much of the silicon unreacted.

## Conclusions

Equilibrium thermodynamics has been used to study the potential reactions in the $Al/Al_2O_3$ system in the presence of small quantities of Mg and Si heated to 1453K in air. The predictions have been made using the MTDATA program and have been compared with the compounds formed during the growth of Al–$Al_2O_3$ composites by the DIMOX™ process using externally doped ultra-pure aluminium and a particulate alumina preform.

In particular it has been shown both experimentally and theoretically that the amount of oxygen present at any time in any region of the reaction system is crucial in determining whether aluminium oxide or nitride is formed in that region at that time. The success of the thermodynamic calculations shows that although the calculations cannot capture all the kinetic aspects of the DIMOX™ reaction system they can be used to predict the compounds which may be formed both during and on completion of the reaction process.

## Acknowledgements

The authors' thanks are due to the National Physical Laboratory, Teddington, Middlesex TW11 0LW for access to MTDATA and for stimulating discussions with their staff.

## References

1. M. S. Newkirk, H. D. Lesher, D. R. White, C. R. Kennedy, A. W. Urquhart and T. D. Claar: *Ceram. Eng. Sci. Proc.*, 1987, **8**(7-8), 879–885.
2. M. S. Newkirk, A. W. Urquhart and H. R. Wicker: *J. Mater. Res.*, 1986, **1**, 81–89.
3. M. K. Aghajanian, M. H. Macmillan, C. R. Kennedy, S. J. Luczcz and R. Roy: *J. Mater. Sci.*, 1989, **24**, 658–670.
4. O. Salas, H. Ni, V. Jayaram, K. C. Vlack, C. G. Levi and R. Mehrabian: *J. Mater. Res.*, 1991, **6**, 1964–1981.
5. K. C. Vlack, O. Salas, H. Ni, V. Jarayam, C. G. Levi and R. Mehrabian: *J. Mater. Res.*, 1991, **6**, 1982–1995.
6. S. Antolin, A. S. Nagelberg and D. K. Creber: *J. Am. Ceram. Soc.*, 1992, **75**, 447–454.
7. A. S. Nagelberg: *J. Mater. Res.*, 1992, **7**, 265–268.
8. R. H. Davies, A. T. Dinsdale, T. G. Chart, T. I. Barry and M. H. Rand: *High Temperature Science*, 1990, **26**, 251–262.
9. I. Ansara and B. Sundman, "Computer Handling and Dissemination of Data', *Proceedings of the 10th CODATA International Conference*, Amsterdam, P. Glaeser, ed., Elsevier, 1987, 154.
10. R. H. Davies, A. T. Dinsdale, S. M. Hodson, J. A. Gisby, N. J. Pugh, T. I. Barry and T. G. Chart: 'MTDATA - the NPL Databank for Metallurgical Thermochemistry' in *User Aspects of Phase Diagrams*, Institute of Metals, London, 1991, pp140–152.

# Activity Coefficients of Oxygen in Liquid Antimony, Tellurium and Antimony–Tellurium Alloys

B. Onderka and K. Fitzner

*Institute of Metallurgy and Materials Science, Polish Academy of Sciences,*
*25 Reymonta Street, 30-059 Krakow, Poland*

## Abstract

The activity of oxygen occurring at low concentrations in liquid Sb, Te, and Sb–Te solutions was studied by means of the galvanic cells with zirconia electrolytes:

$$W \text{ or } Re, \mathbf{O} \text{ in } Sb_x Te_{1-x} \mid\mid O^{-2} \mid\mid air, Pt$$

The coulometric titration technique was employed. The Gibbs free energy of the solution of oxygen in liquid antimony and tellurium may be expressed as:

$$\Delta G^0_{O\,(Sb)} = -129\,075 + 11.045\,T \qquad [\,J\,(g\text{ atom }O)^{-1}\,]$$

and

$$\Delta G^0_{O\,(Te)} = -96\,060 + 38.957\,T \qquad [\,J\,(g\text{ atom }O)^{-1}\,]$$

respectively. The activity coefficients of oxygen were also determined over the whole alloy composition range as a function of the temperature in the interval 923–1123 K. Experimental results were compared with calculated values based on several theoretical models.

## Introduction

The application of electrochemical methods based on the performance of solid oxide electrolytes conducting with oxygen ions enabled the oxygen chemical potential to be measured in a number of high temperature systems. It was Alcock and Belford who, using this technique, first determined oxygen activity in liquid metals, namely lead and tin.[1,2] Since their first coulometric titration experiments numerous investigations of oxygen activity in liquid metals and their alloys have been carried out.[3]

In this laboratory we successfully measured the interaction coefficients between oxygen and arsenic as well as oxygen and phosphorus in several III–V systems.[4–7] However, the range of alloy compositions covered by the experiments was limited by a steep rise of liquidus in systems like Ga–As or Ga–P, for example. Thus, to follow the variation of oxygen activity with the alloy composition in systems

175

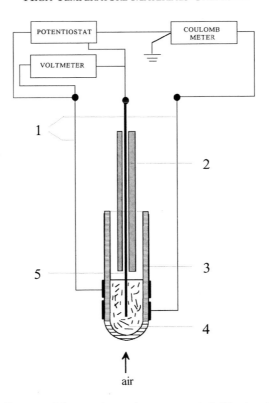

**Fig. 1** *Schematic diagram of the experimental arrangement. 1, Pt wires; 2, alumina tube; 3, zirconia electrolyte; 4, working electrode; 5, tungsten or Kanthal-rhenium wire*

exhibiting semiconducting, solid compounds, other binary systems had to be considered. Therefore, we decided to measure oxygen activity in the Sb–Te binary system. In this case, the oxygen chemical potential can be determined by coulometric titration experiments in the whole alloy concentration range.

## Experimental Details

Antimony (purity, 99.999%) was supplied by Unitra Cemi (Poland) while tellurium (purity, 99.9999%) was obtained from ASARCO (USA). Solid electrolyte tubes, closed at one end, and made of either calcia- or yttria-stabilised zirconia (outside diameter 8 mm and inside diameter 5 mm) were obtained from Friedrichsfeld GmbH (Germany). Respective alloys were prepared by melting proper amounts of pure metals inside the electrolyte tube.

A coulometric titration technique was employed for the determination of oxygen activity in liquid alloys. The experimental method and procedure have been described previously.[4-7] A schematic diagram of the experimental arrangement of the cell:

Pt, air $||$ $O^{-2}$ $||$ O in Sb, Te or Sb-Te, W or Re

is presented in Fig. 1.

Titrations were carried out in the temperature range 673–823 K for pure tellurium, 973–1173 K for pure antimony, and 773–1123 K for Sb–Te solutions. The tube of solid electrolyte contained ca. 3 g of metallic alloy of the chosen composition. The outer part of the solid electrolyte tube coated with platinum paste worked as an air reference electrode and was connected to the electric system with a platinum wire. The electric circuit contained a potentiostat model 550 and a charge meter (integrator) model 721, both produced by AMEL (Italy), and the digital voltmeter 197A made by Keithly (USA). Purified argon was allowed through the cell just above the surface of the liquid metal. Either tungsten wire or Kanthal A-1 wire with a welded rhenium tip acted as electric contacts with the liquid metal electrode.

After the chosen temperature had been reached and the equilibrium electromotive force (emf) of the cell $E_1$ was recorded, the preselected additional potential $\Delta E$ was applied by the potentiostat. The resulting current passed through the cell in such a direction that oxygen was pumped out of the metal. The decrease in oxygen concentration resulted in an increase in the emf of the cell. The final emf value $E_2$ and the electric charge $Q_{ion}$ were recorded. The experimental run was repeated several times at the same temperature, then the temperature was changed.

## Results

Activity coefficients of oxygen in liquid Sb, Te and Sb–Te alloys were calculated from the relationship:

$$f_O = p_{O_2}^{\frac{1}{2}} / C_{O(1)} \tag{1}$$

where $p_{O_2}$ is directly related to the emf through the equation:

$$E_1 = (RT/4F) \cdot \ln (0.21/p_{O_2}) \tag{2}$$

and the oxygen concentration $C_{O(1)}$ can be obtained from the following two equations:

$$C_{O(1)} - C_{O(2)} = 100 \, (M/W) \cdot (Q_{ion}/2F) \tag{3}$$

and

$$E_2 - E_1 = (RT/2F) \cdot \ln (C_{O(1)} / C_{O(2)}) \tag{4}$$

assuming Henry's Law obeys over the oxygen solubility range considered. $\Delta E = E_2 - E_1$ is an imposed potential difference (in milivolts), the oxygen concentration $C_O$ at the beginning (1) and at the end (2) of the pump-out experiment is expressed in atomic %, $M$ is the atomic weight of the liquid metal, $W$ is the mass of liquid metal, $Q_{ion}$ is the quantity of electricity due to ionic current (in Cb), $F$ is Faraday's constant and $R$ is the gas constant.

| $X_{Te}$ | A | B | $r$ | MSE |
|---|---|---|---|---|
| 0.00 | $-15524.29 \pm 725.23$ | $1.32853 \pm 2.01016$ | 0.9728 | 0.18 |
| 0.10 | $-24632.22 \pm 1141.82$ | $8.37568 \pm 1.06755$ | 0.9634 | 0.24 |
| 0.20 | $-25476.57 \pm 1095.24$ | $9.06254 \pm 1.02314$ | 0.9700 | 0.24 |
| 0.30 | $-31200.22 \pm 1237.23$ | $13.13464 \pm 1.15550$ | 0.9736 | 0.26 |
| 0.40 | $-25075.74 \pm 835.13$ | $9.43791 \pm 077990$ | 0.9854 | 0.15 |
| 0.50 | $-16249.62 \pm 1599.39$ | $1.96404 \pm 1.47802$ | 0.8938 | 0.25 |
| 0.60 | $-15195.28 \pm 404.44$ | $0.77804 \pm 0.37765$ | 0.9889 | 0.08 |
| 0.70 | $-19858.66 \pm 641.26$ | $5.22456 \pm 0.63873$ | 0.9635 | 0.28 |
| 0.80 | $-14483.62 \pm 445.89$ | $0.13033 \pm 0.44914$ | 0.9701 | 0.20 |
| 0.90 | $-14454.04 \pm 276.94$ | $1.96114 \pm 0.33399$ | 0.9883 | 0.14 |
| 1.00 | $-11554.57 \pm 585.27$ | $4.68588 \pm 0.70553$ | 0.9545 | 0.27 |

**Table 1** *Results of coulometric titration experiments; ln $f_O$ = A/T + B, r-correlation coefficient; MSE mean square error.*

The activity coefficients of oxygen in dilute Sb, Te and Sb-Te liquid alloys determined in this study in the temperature range 773–1173 K are given in Table 1 and Fig. 2 (a – k).

For ln $f_O$ calculation, oxygen concentration $C_{O(1)}$ and oxygen partial pressure obtained from the emf $E_1$ were used. The least-squares method was applied to experimental data in order to derive ln $f_O$ against reciprocal temperature dependence.

## Discussion

The results obtained for the free energy change of the reaction of dissolution:

$$1/2\, O_2 \Leftrightarrow O \tag{5}$$

in liquid antimony and tellurium can be expressed by the oxygen activity coefficients reported in Table 1:

$$\Delta G^0_{O\,(Sb)} = RT \ln f_{O\,(Sb)} = -129\,075.4 + 11.0449\, T \quad [\text{J/g.atom O}] \tag{6}$$

and

$$\Delta G^0_{O\,(Te)} = RT \ln f_{O\,(Te)} = -96\,060.2 + 38.9566\, T \quad [\text{J/g.atom O}] \tag{7}$$

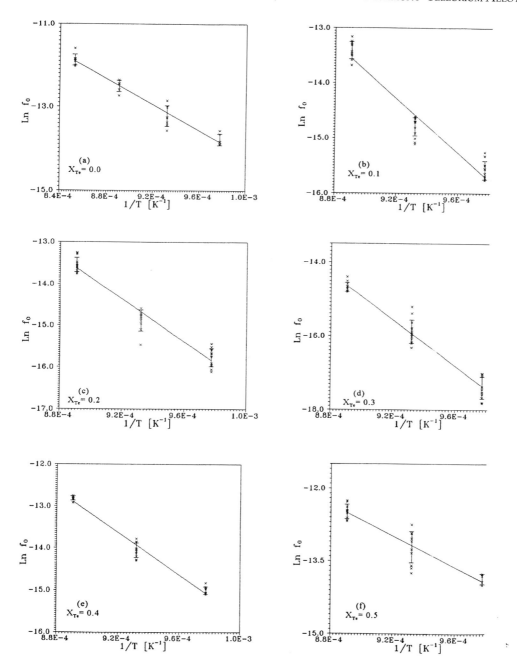

**Fig. 2 a-f** *Experimental results; straight lines correspond to ln* $f_O$ = A/T + B *equations with parameters A and B given in Table 1.*

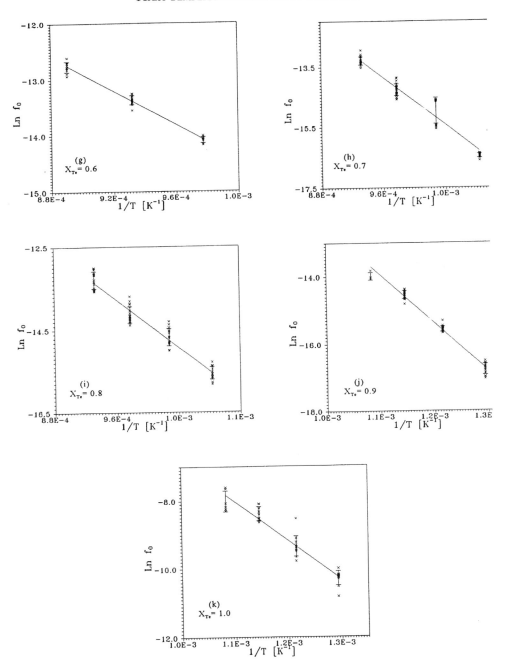

**Fig. 2 g-k** *Experimental results; straight lines correspond to ln* $f_O = A/T + B$ *equations with parameters A and B given in Table 1.*

where the reference state for dissolved oxygen is an infinitely dilute solution in which the activity is equal to atomic percent.

The results obtained are in good agreement (within 5 kJ for Sb and 10 kJ for Te) with the data reported in the literature.[3]

Similarly, activity coefficients of oxygen in liquid antimony–tellurium alloys were determined from the experimental data gathered in Table 1, and they are shown in Fig. 3. For the description of the composition dependence of the activity coefficient, we have chosen a set of data determined at 1073 K, i.e. in the middle of the temperature range covered by our experiments.

Two simple models were used to describe these results; Jacob[N]Alcock's model:[8]

$$\frac{1}{[f_O]^{\frac{1}{n}}} = (1 - X_{Te}) \left\{ \frac{\gamma^\alpha_{Sb}}{[f_{O(Sb)}]^{\frac{1}{n}}} \right\} + X_{Te} \left\{ \frac{\gamma^\alpha_{Te}}{[f_{O(Te)}]^{\frac{1}{n}}} \right\} \qquad (8)$$

and Wagner's model:[9]

$$\frac{1}{f_O} = \sum_{i=0}^{i=Z} \frac{Z!}{(Z-i)! \, 2x} \left\{ \frac{(1-X_{Te})}{[f_{O(Sb)}]^{\frac{1}{z}}} \right\}^{Z-i} \left\{ \frac{X_{Te}}{[f_{O(Te)}]^{\frac{1}{z}}} \right\}^i \exp\left[ \frac{(Z-i) \, i}{2\,RT} \right] \quad h \quad (8)$$

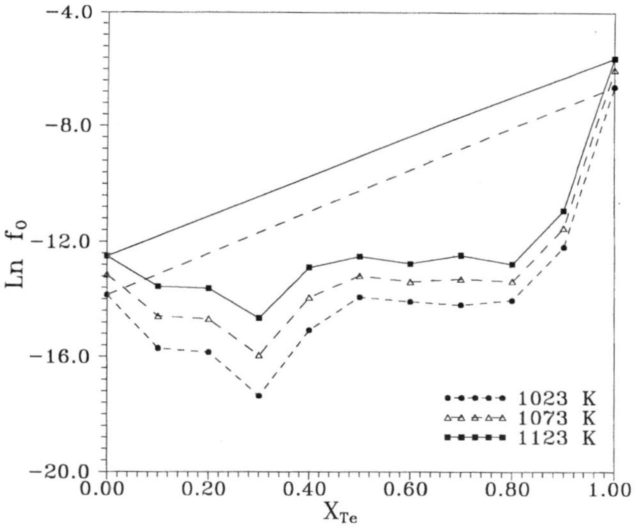

Fig. 3 *Activity coefficient of oxygen in liquid Sb–Te alloys as a function of temperature and composition*

The plots of $\ln f_O$ vs. the alloy composition obtained from eqs. (8) and (9) are compared in Fig. 4 with the experimental data. The values of $\ln f_O$ of pure antimony and pure tellurium are obtained from eqs. (6) and (7), while activity coefficients of antimony and tellurium in the liquid Sb–Te solution have been determined recently in our laboratory.[10] Wagner's model was correlated with parameter **h** equal to 10502.3 J/g.atom. It is clear that both models do not reproduce the dependence shown by the experimental data, even though Wagner's model gives a better fit.

This is not surprising since both models use constant parameters, which are either assumed ($n = 4$ and $\alpha = 1/2$) or obtained by fitting the equation to the experimental data ($h$). However, investigations of the physical properties of liquid Sb–Te alloys, namely electrical conductivity, viscosity, and density,[11–14] indicate the existence of a residual structure in the liquid with the maximum volume portion of associates around the $Sb_2Te_3$ composition. Moreover, a sudden rise in electrical conductivity[15] observed at about $X_{Te} = 0.25$ indicates faster 'metallisation' of the liquid in this part of the system. It is clear that the short range order in the liquid may vary significantly from pure antimony to pure tellurium. Since parameters $n$ and $\alpha$ in eq. (8) are defined by a number of bonds between atoms, they must be a function of a local structure, which apparently depends on the alloy concentration.

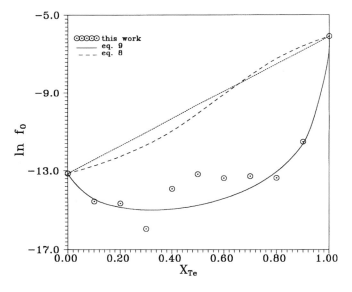

**Fig. 4** *Comparison of ln f$_O$ values determined experimentally at 1073 K and deduced from theoretical models.*

Consequently, one can come to the conclusion that these parameters can not be kept constant but should vary with the alloy composition for a large group of systems with a compound-forming tendency. Thus, while the rigidity of Wagner's model has been partially removed by introducing the parameters $h$ dependent on the composition of the solvation shell,[16, 17] based on the quasichemical approach, Jacob-Alcock's model still has to be improved.

## Acknowledgement

This work was supported by the State Committee for Scientific Research under Grant No. 308 669 101.

## References

1. C. B. Alcock and T. N. Belford: *Trans. Faraday. Soc.*, 1964, **60**, 822.
2. T. N. Belford and C. B. Alcock: *Trans. Faraday. Soc.*, 1965, **61**, 443.
3. Y. A. Chang, K. Fitzner and Min-Xian Zhang: *Progr. Materials Sci.*, 1988, **32** (2/3), 97.
4. J. Wypartowicz and K. Fitzner: *J. Less-Common Met.*, 1987, **128**, 91.
5. J. Wypartowicz and K. Fitzner: *J. Less-Common Met.*, 1988, **138**, 289.
6. J. Wypartowicz and K. Fitzner: *J. Less-Common Met.*, 1990, **159**, 35.
7. B. Onderka, J. Wypartowicz and K. Fitzner: *Arch. Met.*, 1991, **36**, 5.
8. K. T. Jacob and C. B. Alcock: *Acta Met.*, 1972, **20**, 221.
9. C. Wagner: *Acta Met.*, 1973, **21**, 1297.
10. B. Onderka, K. Fitzner: unpublished report, 1993.
11. V. M. Glazov, A. N. Krestovnikov and N. N. Glagoleva: *Izv. AN SSSR, Neorg. Mat.*, 1966, **3**, 453.
12. V. N. Glazov, A. N. Krestovnikov, N. N. Glagoleva and S. B. Evgenyev: *Izv. AN SSSR, Neorg. Mat.*, 1966, **8**, 1477.
13. R. Blakeway: *Phil. Mag.*, 1969, **20**, 865.
14. O. N. Muststsa, G. M. Zagorowsky and A. A. Velikanov: *Izv. AN SSSR Neorg. Mat.*, 1970, **7**, 453.
15. G. F. Gubska and I. V. Evfimovsky: *Russ. J. Neorg. Chem.*, 1962, **7**, 1615.
16. T. Chiang and Y. A. Chang: *Metall. Trans. B.* 1976, **7B**, 453.
17. R. Schmid, J-C Lin and Y. A. Chang: *Z. Metallkde.*, 1984, **75**, 730.

# Control of Inclusions in Steel

R. J. HAWKINS

*British Steel Technical, Teesside Laboratories, P.O. Box 11, Grangetown, Middlesborough, Cleveland, TS6 6UB, UK.*

## Abstract

It is not yet a practical proposition to make inclusion-free steel on a tonnage scale, but it is possible to control the amounts, size, composition and distribution of inclusions to minimise their detrimental effects on product properties. Improved thermochemical models, advanced physical modelling techniques, the development of on-line sensors, the application of computational models and improvements in computing techniques are all part of the armoury available to the modern steelmaker to ensure that a consistent high quality steel can be produced.

Some aspects of the control of inclusions in steel are presented involving mathematical modelling and analytical developments. The engineering of inclusion composition through the use of computational models and preliminary practical experiments to achieve compositional control in the tundish are described. The modelling of flows in continuous casting moulds is discussed including methods of investigating submerged entry nozzle blockage. Computer applications in spectrometry for the rapid determination of steel cleanness are briefly described.

Further developments of thermochemical models for slags are required and data on thermophysical properties, particularly at liquid metal/slag interfaces, will be required for accurate modelling of inclusion behaviour. Lasers for *in situ* analysis of liquid metal are at an advanced stage of development and new methods of optical emission spectrometry present good prospects for rapid and accurate determination of steel cleanness.

## Introduction

As part of the continuing effort to obtain high quality steel products with consistent properties, it is necessary to use a wide range of processing techniques to control the amounts, composition, size and distribution of non-metallic inclusions (NMIs). The detrimental effects of NMIs in steel products can result in operational problems and reduced efficiencies in forming and machining processes; or, perhaps, the defects associated with NMIs may result in a requirement to over-design a component to allow for variability in creep, fatigue or stress corrosion properties. Broadly, the NMIs originate from entrained slags, fluxes or eroded refractory material (exogenous inclusions). Precipitation usually takes place on exogenous inclusions changing their composition or forming shells around them. These reactions lead to heterogeneity in the inclusion population.

Although physical removal of NMIs by filtration has been successful on a small scale, it is not possible to filter steel on a tonnage scale to produce an inclusion-free product. In any case, only the exogenous inclusions could be removed in this way and precipitation of solutes, reoxidation and reaction with tundish or mould fluxes would always ensure that some residual NMI population was present.

The composition of the original exogenous inclusions and the amounts of alloy or deoxidant additions affect the physical properties of the NMIs during casting and subsequent processing. Generally, small spherical inclusions are less detrimental than acicular shapes or agglomerates and the control of the liquidus temperature of the inclusion is one way of ensuring that globular shapes are obtained as the inclusions are incorporated into the solidifying mushy zone in the continuous casting strand. Depending on the steel type, there may also be a need to consider the properties of the inclusion at lower temperatures since further transformations of the oxides, sulphides, carbides, nitrides etc., can take place during the processing of the solid steel. Transformations that produce hard precipitates such as alumina or CaS lead to excessive wear of ceramic tools during high speed machining. On the other hand, the presences of low melting point phases such as MnS can help to lubricate the cutting tool. The interaction of these molten phases with the alumina or carbide tool tip then becomes another factor in the engineering of inclusion type.

Several comprehensive papers on inclusion control in steelmaking have been written,[1-4] the publications by Gatellier *et al.* being particularly recommended. The purpose of this paper is not to discuss inclusion chemistry in depth but to report on selected developments in the use of computer models, analytical techniques and mathematical modelling methods which are used within British Steel to assist the steelmaker in the quest to achieve and sustain high quality product output.

## Inclusion Engineering

IRSID has been very active in developing computational models for slag systems[5] which, when combined with thermochemical models for iron solutions, enable the equilibrium phases in the system Fe–Si–P–Al–Cr–Mg–Mn–F–S–O to be calculated. Thus, the compositional changes of exogenous NMIs and the precipitation of new indigenous inclusions during the casting and solidification of the steel can be investigated. Careful work on verifying these models has been carried out by separating the inclusions by non-aqueous dissolution of the steel and analysing the residual insoluble material.[1] Although a number of alternative slag models could be used, with the associate species model being a strong contender,[6] a modified Kapoor–Frohberg (K–F) model has proved to be quite robust; and this IRSID model[5] is presently used by British Steel for calculating slag and inclusion crystallisation paths.

During secondary steelmaking operations, stirring is sufficient to entrain some of the top slag and it is reasonable to assume that equilibrium is closely approached between the slag, inclusions and steel. The calculation to be undertaken in the inclusion engineering exercise can only give a broad indication of the target metal and slag specifications since the initial exogenous inclusion population will have a range of compositions; and subsequent trimming requires an assumption of reproducible recoveries of alloys (which may be difficult to achieve with volatile elements such as calcium). Fortunately, evidence from plant trials indicates that the bulk of the inclusion population in the tundish has a relatively homogeneous composition for an aluminium killed steel and can be related to the composition of the slag phase in the ladle at the end of treatment. (For a low aluminium steel, at Al concentrations of a few ppm, the inclusion population arising from manganese and silicon deoxidation has a more variable composition).

For the engineering of inclusion composition, the modified K–F model is combined with a model for multicomponent iron solutions describing the activity of solutes using first and second order interactions coefficients. All of the elements in the slag-metal system are partitioned between the iron solution and the liquid and solid oxide phases. The K–F utilises a free energy minimisation method to determine the slag constitution which is assumed to be an agglomeration of cells of the type $M - O - M$ or $M - O - M^1$ where M and $M^1$ are cations. The cells have assigned values for their free energy of formation and there are parameters representing the energies of interactions between cells. When the global minimum for the integral free energy of mixing of the cells has been determined, a numerical partial differentiation is performed to determine partial molar free energies for the constituent (liquid) oxides. Hence, the activities of the oxides can be calculated and used in conjunction with the activities of the appropriate solutes in the steel to determine the activity of oxygen dissolved in steel, $a_o$. The program iterates between slag and metal models, subject to the constraint of mass conservation, until equilibrium is reached. Calculation is complete when:

$$(a_O)_{Al} = (a_O)_{Si} = (a_O)_{Mn} \text{ etc}$$

$$\text{where } (a_O)_M = \left( \frac{K_M a_{MOx}}{a_M} \right)^{1/x}$$

for solute M (i.e. Al, Si, Mn etc), $x$ = (cation valancy)/2 and $K_M$ is the appropriate

**Metal**

| Al | Mn | Si | Temperature (°C) |
|----|----|----|------------------|
| 0.03 | 1.07 | 0.16 | 1630 |
| 0.05 | 1.26 | 0.19 | 1523 |

**Inclusions (50ppm)**

| Initial slag weight (t/100 t steel) | Inclusion composition | | | | Inclusion liquidus temperature (°C) |
|---|---|---|---|---|---|
| | SiO$_2$ | Al$_2$O$_3$ | CaO | MgO | |
| 0.05 | 1.1 | 54.3 | 41.2 | 2.9 | 1479 |
| 0.5 | 1.0 | 55.1 | 40.0 | 3.4 | 1483 |
| 1.0 | 0.9 | 55.4 | 39.7 | 3.5 | 1485 |

10ppm Ca added

**Table 1** *Effects of initial slag/metal equilibration on inclusion properties.*

equilibrium constant.

A decision is required on the choice of the relative amounts of metal and slag or inclusions for the calculation. During the ladle treatment, some of the top slag will be emulsified into the steel and most of it will float out again. Table 1 shows the effects of equilibrating different amounts of slag at 1630°C to produce a starting population of exogenous inclusions consisting of a residual fine dispersion of slag droplets, in an off-shore plate grade steel, which will undergo further transformation as the inclusion-liquid metal system cools down during casting. The intuitive choice for the amount of exogenous inclusions present in the steel corresponds to the analysed total oxygen expressed as a quantity of oxide. This will vary according to plant practice, the type of steel and the deoxidation procedures. In this example, 50 ppm by weight of inclusions (i.e. 50 g per tonne of steel) was chosen. It can be seen that the final inclusion compositions and liquidus temperatures (1479–1485°C) are relatively insensitive to the amount of slag originally equilibrated with the metal.

| Inclusion/ metal ratio (ppm) | Inclusion composition | | | | Inclusion liquidus temperature (°C) |
|---|---|---|---|---|---|
| | $SiO_2$ | $Al_2O_3$ | CaO | MgO | |
| 20 | 1.15 | 54.0 | 41.7 | 2.7 | 1473 |
| 50 | 0.9 | 55.4 | 39.7 | 3.5 | 1485 |
| 100 | 0.9 | 56.7 | 38.0 | 3.9 | 1506 |
| 120 | 0.9 | 57.1 | 37.6 | 4.0 | 1514 |
| 150 | 0.8 | 57.8 | 36.9 | 4.2 | 1526 |

Initial equilibration at 1630°C
Metal: 0.05% Al, 1.26% Mn, 0.19% Si, 1523°C
10 ppm Ca added

**Table 2** *Effect of retained inclusion concentration on inclusion properties.*

The liquidus temperature of the steel is 1523°C, so the inclusions should be globular. Table 2 shows the effects of attempting to carry out a standard calcium treatment (10 ppm addition) without proper knowledge of the amounts of exogenous inclusions. As the relative amounts of inclusions are increased, the calcium present is insufficient to achieve full modification to retain liquid inclusions down to the liquidus temperature of the steel; and clustering of solid or semi-solid inclusions could take place, leading to poor product quality and perhaps could also interfere with the casting operation by causing blockages. It is important to ensure that castability is maintained, i.e. the liquid steel is not too erosive towards the refractory ware, nor on the other hand, likely to cause blockage by deposition of solid inclusions. The former situation arises when excessive amounts of calcium are present which can react with alumina refractories, in particular, and thus destroy the geometry of the casting nozzle.

## Compositional Control in the Tundish

Having indicated that thermochemical models may be used to ensure that an acceptable inclusion population can be obtained in the steel product, it is necessary to devise reliable methods of monitoring and achieving the correct steel specification on plant. The continuous monitoring of iron and steel composition and the use of various types of sensors have been the subject of research within British Steel for many years and one approach has been to use lasers for *in situu* analysis. Laser analysis can be applied to solid or liquid metal and also to non-conductors. Thus, in principle, *in situu* analysis of both metal and slag can be carried out during ironmaking, oxygen steelmaking and secondary steelmaking operations.

Two principles of laser analysis have been studied.[7] The first involves the interaction of a focused laser beam with the liquid metal surface when material is ablated and forms a plasma.[8] Energy densities of around $10^{10}$ watts cm$^{-2}$ are required. The light emitted by the plasma can be analysed spectrometrically to give the concentrations of C, Si, S, Mn, Al, Ca etc. With proper time-gating of laser events, analytical accuracy for elements of high atomic number is comparable to that achieved in conventional laboratory-based optical emission spectrometry (OES) used for routine quality control. The second method of analysis, which has not yet been extensively investigated, is to make use of the smoke formed by the condensation of ablated material. This smoke can be ducted through tubing to an inductively coupled spectrometer and has been analysed in a laboratory up to 80 m distance from the vessel containing the liquid steel.[7]

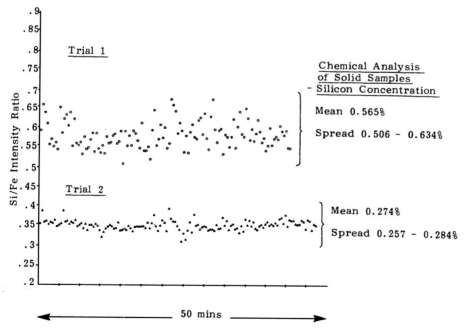

**Fig. 1** In situ *laser analysis for silicon in iron in a blast furnace runner.*

An *in situ* laser has been used in the past on the blast furnace at Redcar to monitor silicon (Fig. 1). The equipment used a 700 mJ Nd YAG laser firing at 10 Hz with an optical fibre linking the probe to a spectrometer. Each point in the figure corresponds to a 10 s integration, i.e. 100 laser events. The analyser operated continuously for about an hour before the trial was terminated prior to the end of tapping.

Current work at Teesside Laboratories on compositional control in the tundish has been undertaken using a laser analyser. The main incentives for this study are related to (i) meeting the narrower compositional and quality ranges now demanded for steel; (ii) controlling or countering interactions with air, ladle slags, tundish and mould fluxes; (iii) engineering inclusion compositions and (iv) maintaining castability during long sequence casting. The latter mainly implies controlling Al and Ca contents. There may also be financial benefits of around 50p/tonne in transferring some of the compositional control from the secondary steelmaking vessel to the tundish by trimming to the lower specifications limits for C, Si, Mn, Al and Ca. Further savings may also be made by improving yields at a change of grade when approximately 25% of the tundish contents may be discarded during the changeover (i.e. 12–15 tonnes) when it could be trimmed to match a succeeding grade.

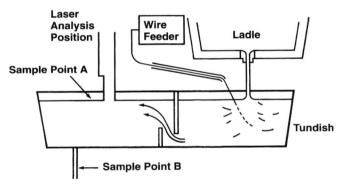

**Fig. 2** *Schematic of pilot plant for compositional trimming in the tundish.*

A schematic of the pilot plant tundish showing the laser and alloy wire feeder for compositional trimming is illustrated in Fig. 2. The output from the laser analyser in response to Si and Mn additions is shown in Fig. 3; and this output has been linked directly to a PC in order to control the speed of the alloy wire feed to maintain a constant steel composition when required. The consistency and suitability of the analysis were determined for control action based on 20 integrations of 5 laser shots (10 s process time) when the rolling average over 200 integrations showed a relative standard deviation (RSD) of about 3% for Si and 2% for Mn where:

$$\text{RSD} = \frac{\text{s.d. of 10 rolling means}}{\text{mean of 10 rolling means}}$$

This compares with a RSD of 1% for Si and Mn for solid samples analysed by OES in a typical analytical laboratory. Of course, the RSD, for the laser analyser can be

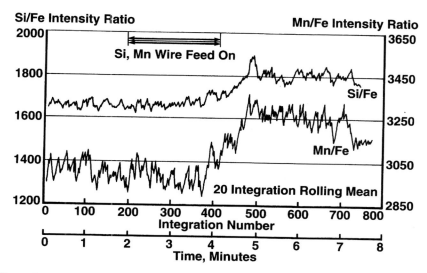

**Fig. 3** *Response of laser analyser to 0.05 wt% Si and Mn alloy wire additions.*

further improved by prolonging the process time; but the quoted precision is sufficient for process control purposes.

Further developments in laser analysis are taking place. For instance, by carefully selecting very high-grade optical fibres, it is possible to transmit UV light emitted by carbon in the plasma event. However, a recently constructed commercial prototype laser analyser designed by British Steel achieves higher sensitivity by incorporating two spectrometers. Direct viewing of the laser event enables carbon and sulphur to be analysed and, with further developments in spectrometer technology and signal processing, it should be possible to determine phosphorus contents continuously too. It is unlikely that nitrogen contents can be determined; and oxygen and hydrogen cannot be analysed by laser spectrometry. Fortunately, other established and developing methods for continuous monitoring exist for the latter two elements.

Notwithstanding the use of tundishes for compositional control, it is usual to design conventional tundishes to allow flotation of inclusions. This has been well covered in the technical literature[9] (this is a reference to recent work which is quite typical – and also contains further references) where computational fluid mechanics is widely used to predict tundish performance with respect to inclusion flotation and temperature distribution. However, transient effects, especially those associated with ladle changeover, often require physical modelling studies in addition to, or instead of, computational models in order to resolve design and operational problems in a cost effective and comprehensive way.

The role of the tundish slag and the refractory surfaces in the tundish in retaining inclusions is still a subject of debate and an interesting contribution to enhancing physical modelling capabilities was made by Hills[10] in devising a wax inclusion analogue of alumina in liquid steel. Rather than repeat previous studies on tundishes, the application of the wax inclusion technique to the design of submerged entry nozzles in continuous casting will be briefly examined.

## Inclusions in the Continuous Casting Mould

Apart from the problems caused by NMIs in the steel product, as stated previously, they can also result in operational problems during the continuous casting process. Alumina inclusions are very prone to forming agglomerations which can build up on the stopper rod and interfere with flow control. They may occasionally break off and flush through as coarse macro-inclusions which could become trapped in the solidifying product. Alumina also deposits and builds up in the bore and the ports of the submerged energy nozzle (SEN) and this can lead to asymmetry in the flow of steel into the mould in the case of slab casting through bifurcated SENs. In the worst cases the bores become blocked. At the other extreme, as mentioned earlier, excessive amounts of calcium, used for inclusion modifications, can lead to SEN erosion which, again, could result in poor distribution of steel into the mould.

A schematic of the flow into a slab mould fed by a bifurcated SEN is shown in Fig. 4. Part of the primary flow into the mould circulates downwards and inclusions already present in the steel as it enters the mould are carried deep down into the

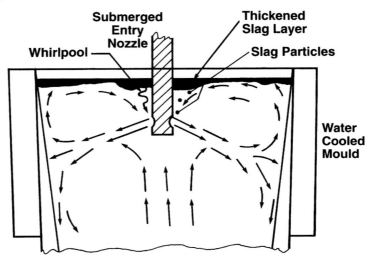

**Fig. 4** *Schematic of the flow of steel in a slab mould fed by a bifurcated submerged entry nozzle.*

sump of liquid steel. Some of these inclusions will be trapped on dendrites and remain in the steel; others will recirculate back towards the surface and may enter the upper vortex and be brought near to the mould flux/steel interface where they could escape from the steel. It is important to maintain steel flow velocities within certain limits in critical parts of the mould. The heat transfer between steel and flux must be sufficient to ensure that the flux melts and to ensure that the surface of the steel is not chilled such that it freezes and plates over. On the other hand, the flow must not be so vigorous as to emulsify and entrain mould flux which could be pulled down into the deeply recirculating vortices.

The conditions at the flux/metal interface are of considerable interest since it can be anticipated that the interfacial tension, the viscosity of the flux, the accumulation of inclusions near to the interface and the assimilation of inclusions resulting in changes in the local flux composition could all be critical in affecting steel cleanness. In many casters, the conditions at the interface are further complicated by argon gas injection into the SEN which is designed to help prevent nozzle blockage. This gas disturbs the interface, may affect the flow pattern (and even reverse it) and, if the gas flow and rate of mould flux additions are not controlled, will clear flux away from the surface leading to reoxidation by air; and, in extreme cases where the flow in the SEN is affected, result in severe asymmetry in the flow patterns. Additionally, when poorly designed SENs are installed, inclusions, particularly titanium carbide, can accumulate in relatively quiescent areas at or near the liquid metal meniscus and be pulled into the solidifying surface layer of steel to give poor surface quality in the product.

   Having described the problems, it is possible to overcome many of them by the use of physical modelling techniques to optimise flow geometry. In particular, the problem of inclusion build-up has been investigated using wax particles to simulate alumina. The wax inclusion generator is shown schematically in Fig. 5 and a

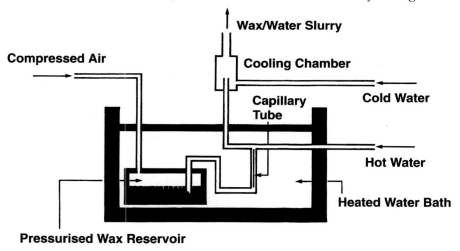

**Fig. 5** *Wax inclusion generator for use with physical models.*

comparison of the model and an actual partially blocked nozzle is shown in Fig. 6. Such techniques are useful in evaluating novel SEN designs and in defining optimum flow control conditions and devices.

   The use of the wax/water analogue shows qualitatively where agglomeration of inclusions could occur; and is particularly useful in providing a practical demonstration of the accumulation of inclusions for the guidance of plant and design engineers. The behaviour of single spherical particles in gravitational or rotational fields, which cause the particles to come together to form agglomerates, can be related between plant and model studies by choosing conditions to ensure Froude number similarity. The relationship between the diameter of alumina particles, $d_{Al_2O_3}$ and the diameter of wax particles, $d_{wax}$, is given by:

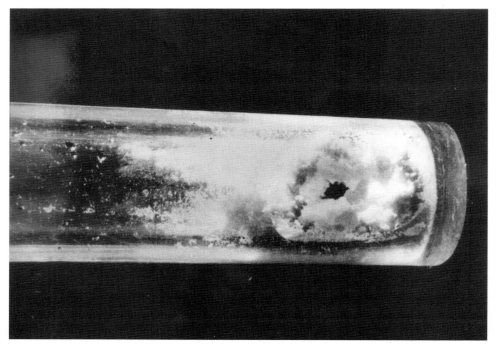

**Fig. 6** *Comparison of wax model simulation and an actual partially blocked nozzle.*

$$d_{Al_2O_3} = 1.46\, S^{\frac{1}{4}} (1 - \rho_{wax})^{\frac{1}{2}}\, d_{wax}$$

where $S$ = model scale and $\rho_{wax}$ = density of wax. Typically, for a half scale model the equivalent alumina particle would be around 0.5 times the wax particle diameter.

Much of the work on producing realistic dynamic computational models of physical and chemical interactions between flux and steel remains to be done. Two, or even three or four phase flow models are required to describe the fate of inclusions and entrained flux. Accurate descriptions are required of interfacial tensions, viscosities, conductivities, activities, liquidus temperatures, etc., for the multicomponent systems $CaO–Na_2O–SiO_2–Al_2O_3–MgO–FeO–MnO–CaF_2$. Assessment work[6] is being carried out at the National Physical Laboratory with the support of a wide range of industries and there is good co-operation across Europe between academic and industrial research institutions. Unfortunately, the present and anticipated requirements for this type of study are not matched by the availability of a broad range of experimental data and validated slag models. Also, on the fluid flow front, turbulence models are probably still inadequate for detailing modelling of small-scale transient events of inclusion capture and re-entrainment at interfaces.

## Rapid Determination of Steel Cleanness

Advances in computer technology for rapid data acquisition and processing as well as for running ever more complex mathematical models is a key feature supporting inclusion engineering, laser analysis, tundish metallurgy and flow modelling. As a final chapter in this short story of inclusions in steel, recent developments in determining the distribution and type of inclusions in the final product will be described. Spectrometer systems have been developed where the emitted light intensities for elements of interest can be stored for each individual spark on an OES operating with a 400 Hz spark source. A project part-funded by European Coal and Steel Community is currently underway at Teesside Laboratories where the element intensity data are combined with positional data during a raster scan of samples of, typically, 100 mm x 100 mm. As the spark strikes the metal it will occasionally vaporise oxides, sulphides and other inclusions as well as vaporising parent metal. Thus, if this information is stored on a PC, it is then possible to produce a compositional map on a macro-scale revealing the distribution of elements or combinations of elements to identify aluminates, silicates etc. Only a linished surface is required and the method can, potentially, replace such time-consuming activities as sulphur printing. Development is still at a relatively early stage, but the method does promise a wealth of data which can be obtained relatively cheaply and quickly in order to diagnose (and correct) problems during casting. Line scans for a clean and a dirty steel are shown in Fig. 7 where it can be seen that, in the dirty steel, there are peaks in aluminium intensities corresponding to the presence of clusters of inclusions. At the moment, there is inadequate control of the location of the arc root and the 'mapping' is only accurate to ±5 mm. However, this still produces an excellent overall assessment of steel cleanness since such a large

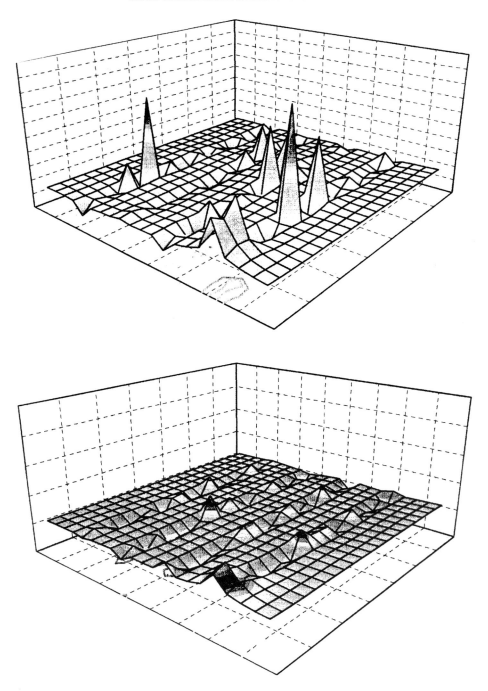

**Fig. 7** *Comparison of line scan analysis for aluminium in (a) dirty steel and (b) clean steel.*

area of steel is sampled compared to metallographic cleanness assessment techniques. Further developments may include the use of micro-electrodes and the use of laser excitation.

## Conclusions

A broad view of selected developments at British Steel Technical, Teesside Laboratories, related to inclusion control has been provided to demonstrate the techniques that are used to ensure that high quality steel products can be manufactured consistently. In all of the areas, ranging from the secondary steelmaking process through tundish metallurgy to the continuous casting mould, further developments of hardware software and theoretical models are anticipated.

1. Slag thermochemical models are not sufficiently accurate in all parts of multicomponent systems and further minor components such as $Na_2O$, $TiO_2$, sulphides, fluorides and phosphates need to be incorporated for calculation of phase equilibria, activities, liquidus and solidus temperatures.

2. Thermophysical properties including interfacial tensions, viscosities and conductivities are required for modelling flux/steel interactions in moulds (and tundishes).

3. Laser analysis is developing rapidly as an acurate method for analysing for all elements in liquid steel on a continuous basis except for hydrogen, nitrogen and oxygen (and possibly phosphorus).

4. The rapid determination of steel cleanness by analysing large areas of steel by optical emission spectrometry has been demonstrated; and macro-scale compositional mapping for quality assurance purposes is being developed.

## Acknowledgements

The author would like to thank Dr R. Baker, Director, Research and Development, for permission to publish this paper. The support of the European Coal and Steel Community for the research activities on laser analysis, tundish metallurgy and the rapid determination of steel cleanness is gratefully acknowledged.

## References

1. C. Gatellier, H. Gaye, J. Lehmann, J. Bellot and M. Moncel: 'Inclusion Control in Low Aluminium Steels,' *La Revue de Metallurgie – CIT*, 1992, 361–369. See also same paper in *Proceedings of 4th International Conference on Clean Steel*, Hungary, 638-651, Institute of Materials, 1992.

2.  C. Gatellier, H. Gaye, J. Lehmann, J. N. Pentoire and P. V. Riboud: 'Physico-chemical Aspect of the Behaviour of Inclusions in Steel,' *Steel Research*, 1993, **64** (1), 87–92.
3.  D. A. R. Kay and J. Jiang: 'Slag Models and Inclusion Engineering in Steelmaking,' *3rd International Conference on Molten Slags and Fluxes*, pp263–269, The Institute of Metals, 1989.
4.  K. W. Lange: 'Thermodynamic and Kinetic Aspects of Secondary Steelmaking Processes,' *International Materials Reviews*, 1988, **33** (2), 53–89.
5.  H. Gaye and D. Coulombet: 'Thermochemical and Kinetic Data on Steelmaking-related Materials,' *Report EUR 9428 FR*, Commission des Communautés Européennes,1985.
6.  T. I. Barry, A. T. Dinsdale and J. A. Gisby: 'Predictive Thermochemistry and Phase Equilibrium of Slags,' *Journal of Metals*, 1993, **45** (4), 32–38.
7.  R. Jowitt: 'Direct Analysis of Liquid Steel by Laser,' *Thirty-eighth Chemists' Conference*, 19-29, British Steel Corporation, Teesside Laboratories, Middlesbrough, 1985.
8.  W. Tanimoto, A. Yamamoto and K. Tsunoyama: *Kawasaki Steel Giho*, 1989, **21** (2), 100–106.
9.  A. K. Sinha and Y. Sahai: 'Mathematical Modelling of Inclusion Transfer and Removal in Continuous Casting Tundishes,' *ISIJ International*, 1993, **33** (5), 556–566.
10. A. W. D. Hills: presented at Conference on *Measurement Techniques in Steelmaking and Casting*, March 1991, Institute of Materials.

# Surface and Interfacial Tension Studies in the Matte/Slag Systems

J.M. Toguri and S.W. Ip

*Department of Metallurgy and Materials Science, University of Toronto, Toronto, Ontario, Canada M5S 1A4*

## Abstract

Surface and interfacial tensions are the underlying driving forces for a wide variety of phenomena in pyrometallurgical processes such as the mechanical entrainment of matte droplets in slag, infiltration and localised corrosion of refractory bricks by matte and slag and electrocapillary motion of matte droplets in slag under an electric field. In order to gain insight into these processes, knowledge of the surface and interfacial tension of these matte/slag systems is a prerequisite.

This paper reviews the surface and interfacial tension of the system Cu-Fe-S, Cu-Ni-S, and Ni-Fe-S together with fayalite slags obtained by using the sessile drop (SD) and/or the maximum bubble pressure (MBP) technique. Based on these data, the entrainment behaviour of matte droplets suspended in slag are evaluated using the filming and flotation coefficients proposed by Minto and Davenport. Infiltration behaviour of matte on various oxides are also discussed.

Electrocapillary motion of matte droplets in slag are considered and the droplet behaviour is interpreted by taking into account the electrocapillary diagram (dependency of interfacial tension with applied potential).

## Introduction

The loss of pay metal to the slag phase due to mechanical entrainment represents a substantial loss in terms of metal recovery. Even with the advancement of various slag cleaning techniques, entrainment is still inevitable. From knowledge of surface and interfacial tension data, this paper looks at the mechanism of metal loss due to mechanical entrainment of matte/metal droplets in slag. A new and novel slag cleaning technique based on the principle of electrocapillarity is also discussed.

Refractory wear contributes significantly to the cost of the metal extraction operation. Infiltration of matte and slag into refractory pores plays an important role in the accelerated degradation of refractory lining since their presence alters both the mechanical and thermal properties of the bricks. In the present study, the infiltration behaviour of matte into various oxide materials are evaluated based on wetting and surface and interfacial tension measurements.

Electrocapillarity is a phenomenon which is responsible for the movement of small liquid droplets (dispersed in a second liquid phase) situated within an electric field. This movement is caused by the presence of a potential gradient across the length of the droplet. As a result, an interfacial tension gradient is present on the droplet surface. This initiates the Marangoni convection within the drop which in

199

turn propels the drop forward. Owing to the speed at which the drops can travel even under a very small electric field, this phenomenon may be utilised to improve the metal recovery in electric slag cleaning furnaces.

## Surface and Interfacial Tension of Matte/Slag Systems

*Experimental*

The sessile drop (SD) and maximum bubble pressure (MBP) technique were adopted as the methods for measuring the surface and interfacial tension of the matte/slag systems. However, at high temperature the SD technique combined with X-ray radiography is the only viable method for measuring interfacial tension. Therefore, all the interfacial tension measurements were performed exclusively using the SD technique.

*Technique*

For the SD technique, an image of the liquid matte drop was obtained by a high temperature X-ray setup. The experimental technique has been discussed in detail elsewhere[1] and thus only a brief description is given here. A simplified schematic of the apparatus is shown in Fig 1.

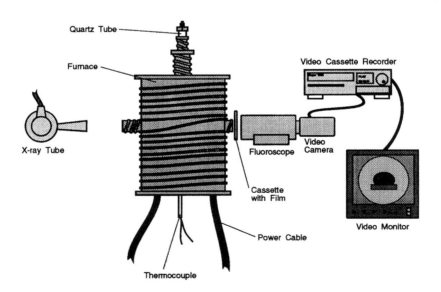

Fig 1. *High temperature X-ray radiographic apparatus.*

For the surface tension experiments, both the FeS and $Ni_3S_2$ were synthesised in the laboratory from their respective elements. Chemical analysis of these sulphides showed the FeS contained 61.3% Fe and 38.7% S, the $Ni_3S_2$, was 69.7% Ni and 30.5% S and the $Cu_2S$ contained 83% Cu and 17% S.

The sample and substrate under investigation were placed into a quartz tube on top of an alumina or boron nitride rod which acted both as support for the sample and weight for the capsule to dampen the pendulum motion of the suspended capsule. The quartz tube was evacuated, sealed and then suspended in the furnace. A schematic of the capsule showing the relative position of the different pieces is shown in Fig 2. In the experiments five different types of substrate materials were used, quartz, alumina, sapphire, magnesia and a high chromium oxide brick (C1215). To perform an experiment a quartz capsule containing the matte sample and oxide substrate was lowered into the furnace via a smaller quartz tube containing a type R (Pt/Pt-13%Rh) thermocouple. An overall arrangement within the reaction tube is illustrated in Fig 3.

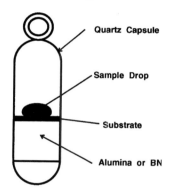

**Fig 2.** *Schematic of quartz capsule used for surface tension experiments.*

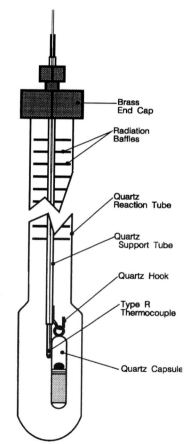

**Fig 3.** *Apparatus arrangement inside the reaction tube for surface tension determination.*

After the location of the sample in the furnace was properly aligned with the X-ray path, the sample was heated to the experimental temperature (1473 K). Events occurring inside the furnace during heating were closely monitored with the X-ray imaging system. Two radiographs were taken after the drop shape was determined to be suitable for surface tension measurements. Subsequently two radiographs were each taken at time steps of approximately 15 min, $^1/_2$ hr, 1 hr, $1^1/_2$ hr and 2 hrs after reaching experimental temperature.

A fayalite type slag was used for all interfacial tension experiments. The slag was prepared synthetically by melting a mixture of iron, silica and hematite at around 1573 K. Chemical analysis of the quenched slag showed 15.3% $Fe_3O_4$, 32.4% $SiO_2$, 51.4% total Fe by mass with trace amount of $Al_2O_3$ (0.28%) and CaO (0.11%). This gave an $Fe/SiO_2$ ratio of 1.58. In preparation for an experiment, a piece of matte weighing about 0.45g was put into the bottom of a high density alumina or magnesia crucible. About 1.5–2g of slag was charged subsequently into the crucible on top of the matte sample. To reduce slag oxidation and chemical reaction between the matte and the slag, the interfacial tension experiments were also performed with sealed quartz capsules. Instead of alumina or boron nitride support, a piece of ~1.5 cm long zirconia rod was used on the bottom of the capsule to support the crucible and act as a weight. Figure 4 is a schematic representation of a quartz

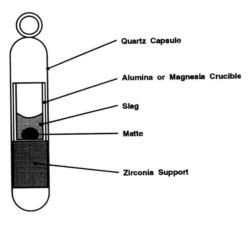

Fig 4. *Schematic of a quartz capsule used for interfacial tension experiments.*

capsule used for interfacial tension measurements. An average interfacial tension experiment lasted only about 15 mins. This short experimental time minimised the amount of dissolved alumina and MgO into the slag. To obtain the drop image, a high resolution DuPont Cronex 10 X-ray film combined with a DuPont Quanta Detail intensifying screen cassette was used. For each experiment, two radiographs were taken at 1–2 min intervals. After each experiment, the samples were cooled similar to the surface tension experiments, The matte and the slag in the crucible were carefully separated and analysed.

To evaluate the surface and interfacial tension from the X-ray images, advanced computer software[2] was used. The profile points from the drop image which are required for the computer calculation are either digitised manually or by computer. Manual digitisation was used for images obtained from interfacial experiments while drops from surface tension experiments were digitised automatically.[3]

## MBP Technique

As for the MBP technique, the single capillary approach was adopted. A schematic diagram of the experimental apparatus is shown in Fig 5. The apparatus consisted of a silicon carbide furnace (1), a gas train (2), an alumina capillary tube (3) with radius in the range of 0.08–0.095 cm, an alumina crucible for containing the melt

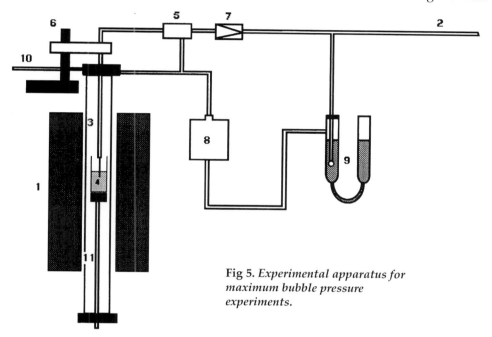

Fig 5. *Experimental apparatus for maximum bubble pressure experiments.*

(4), a pressure transducer (5) for measuring the gas pressure, and a motor drive mechanism (6) for raising and lowering the capillary to a precise depth. The gas bubbling rate in the melt was regulated by a gas overflow arrangement (9) and any pressure fluctuation inside the furnace was stabilised by a buffer vessel (8). Temperature of the sample was monitored by a Pt-Pt/13%Rh thermocouple (11). The thermocouple sheath also served as a support rod for the crucible. To start an experiment, the sample was first placed in the cold zone of the furnace at the bottom of the reaction tube and sealed. Ultra high purity argon was used to purge the tube for 12 hrs while the buffer vessel was closed off from the reaction tube. At the end of the purge, the furnace was brought up to the required experimental temperature and the sample was pushed slowly up to the hot zone. After about 1 hr of heating and stabilising, the capillary was lowered to one of the three chosen immersion depths (0.635, 1.270 and 1.905 cm). The height of the capillary immersion was carefully measured by a cathetometer, The surface of the melt was determined by lowering and raising the capillary in the melt. When the pressure transducer just read zero, the capillary tip location corresponded to the surface of the melt. The path to the buffer vessel was opened before initiation of the gas bubbling. Ultra high purity argon was used as the bubbling gas and the bubbling frequency used was in the range of 30–60 s per bubble. Ten to 15 pressure measurements were obtained for each immersion depth. For each composition, three to four independent experiments were performed, The surface tension data were calculated based on the methodology given by Sugden.[4,5] Further details of the experimental procedure and apparatus are given in references 6, 7 and 8. The three different sulphides, $Cu_2S$, FeS and $Ni_3S_2$, used in MBP experiments were prepared using high purity metals and sulphur. The final sulphur content in these 'master' samples as found to be 20.6%, 37.0% and 26.7% for $Cu_2S$, FeS, and $Ni_3S_2$ respectively. When a mixture of these sulphides was required, it was melted in the experimental furnace prior to each experiment.

*Results and Discussion*

Surface tension of the copper, nickel and iron sulphides, measured as a function of temperature using the MBP technique, are presented in Fig 6. The surface tension

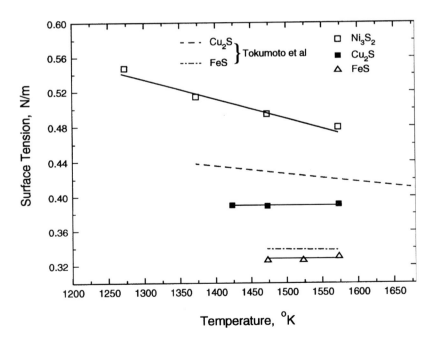

**Fig 6.** *Effect of temperature on the surface tension of some pure sulphides.*

of $Ni_3S_2$ decreases linearly with increasing temperature while FeS and $Cu_2S$ show negligible temperature dependency. The surface tensions of FeS from the present study compare well with Tokumoto et al's[9] data. However, the results for $Cu_2S$ show a deviation of approximately 10%. This difference may be due to possible compositional variations and the difference in the technique used between the two studies. Tokumoto et at used the sessile drop technique together with Bashforth and Adams' table to evaluate surface tension. This approach is susceptible to human bias since only three specific points are selected from the sessile drop profile for surface tension evaluation.

Surface tension of the $Cu_2S$–FeS, $Ni_3S_2$–FeS and $Cu_2S$–$Ni_3S_2$ pseudo-binaries at 1473 K are given in Figs 7 - 9 respectively. The results of the $Cu_2S$–FeS system, Fig 7 (obtained using only the MBP technique), show a non-linear relationship between surface tension and matte composition. The deviation from ideal solution behaviour indicates that FeS is surface active in the system. When the present results are compared with the literature data,[9, 10] the surface tensions obtained in this study show lower values consistently. Nevertheless, all the results showed a similar dependency of surface tension on matte composition. Results of the $Ni_3S_2$–FeS system (Fig 8) show a less dramatic effect of FeS on the surface tension. As indicated in Fig 8, the results obtained using both SD and MBP techniques are in very good agreement. Comparison with Vaisburd et al[11] and Liu et al's[12] data also indicated

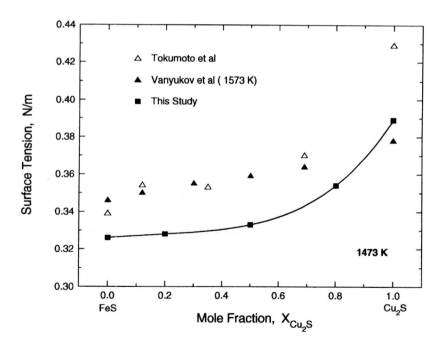

**Fig 7.** *Surface tension isotherm of the Cu–Fe–S system at 1473 K.*

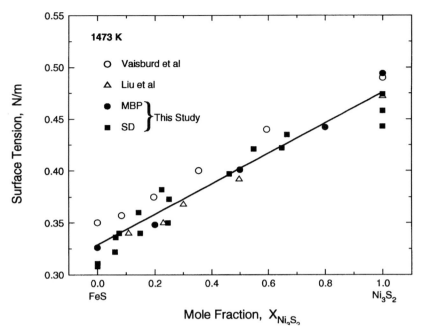

Fig 8. *Surface tension isotherm of the Ni–Fe–S system at 1473 K.*

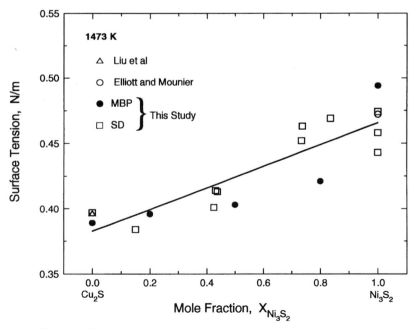

Fig 9. *Surface tension isotherm of the Ni–Cu–S system at 1473 K.*

good agreement. When all the data are combined, a linear relationship between surface tension and composition is observed. However, it should be kept in mind that some of the compositions used in the SD experiments did not fall exactly on the $Ni_3S_2$–FeS pseudo-binary because of either a sulphur deficit or excess in the melt. The surface tension of the $Cu_2S$–$Ni_3S_2$ system (Fig 9) also exhibited a linear relationship with composition.

The surface tension of the Ni–Fe–S system are plotted on a ternary diagram in Fig 10. The iso-tension lines indicate the manner in which changes in either nickel or sulphur content in the matte alter the surface tension of the system. At constant Ni : S ratio, the surface tension of the matte is almost unaltered by the addition of iron to the system.

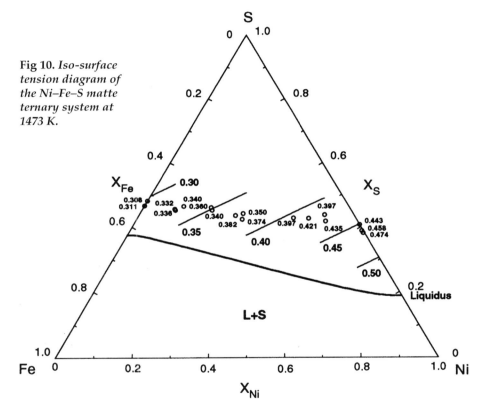

**Fig 10.** *Iso-surface tension diagram of the Ni–Fe–S matte ternary system at 1473 K.*

One of the benefits of the SD technique is that besides surface tension data, the contact angle between the drop and the underlying solid substrate material is obtainable. Fig 11 shows the contact angles between Ni–Fe–S matte drops and various solid oxide substrates. The graph shows that the contact angle between the drop and the substrate increases with increasing nickel sulphide content in the matte regardless of the type of substrate used. Therefore wetting of the solids by matte is favoured by high iron sulphide contents. From Figure 11, the wetting

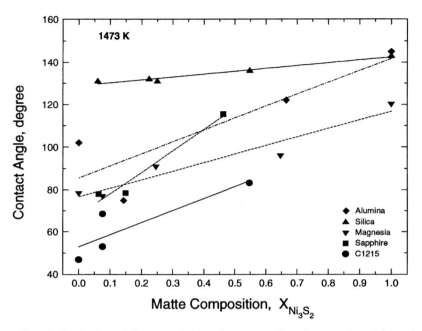

**Fig 11.** *Contact angle between Ni–Fe–S matte and various oxides at 1473K.*

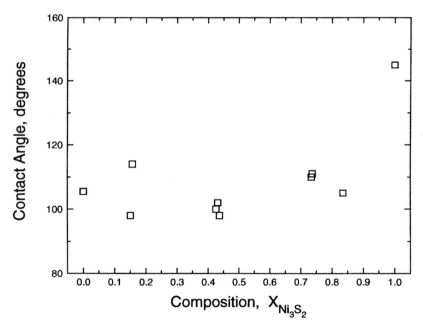

**Fig 12.** *Contact angle between Ni–Cu–S matte and alumina as a function of composition.*

behaviour of the solid oxides by nickel matte can be ranked in order of increasing wettability as follows:

$$SiO_2 < Al_2O_3 \leq Sapphire \leq MgO < C1215$$

Contact angles obtained for the Ni–Cu–S system on alumina substrate are shown in Fig 12. Disregarding the value for pure $Ni_3S_2$, the results show little dependency on matte composition.

Interfacial tension data (SD technique) between nickel mattes and a fayalite slag at 1473 K using both alumina and magnesia crucibles are shown in Fig 13. Very

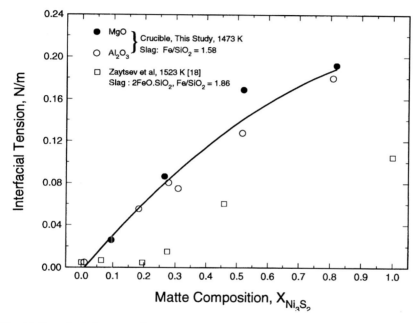

Fig 13. *Interfacial tension between nickel-iron matte and fayalite slag at 1473 K.*

good agreement was obtained between the two sets of results determined from this investigation. When the present results are compared with those obtained by Zaytsev et al[13] their data showed much lower values. Since little information was available on Zaytsev et al's data, it is not possible to assess the observed disagreement which may be due to the difference in density data and slag composition used in the experiments. Nevertheless, results from both studies indicate that the interfacial tension behaviour is non-ideal. Interfacial tension data of the Cu–Fe matte/slag system measured at 1473 K are shown in Fig 14. The data are compared to those reported by Elliott and Mounier[14] and Nakamura et al[15]. Fig 14 shows that there is good agreement amongst the results obtained from the three studies. The data indicate a clear trend of increasing interfacial tension with increasing matte grade. With results similar to those of the Ni–Fe matte/slag system, the interfacial tension value approaches zero as the matte composition approaches that of pure iron sulphide. The low interfacial tension value suggests that the strength of the bonds across the interface are very similar.

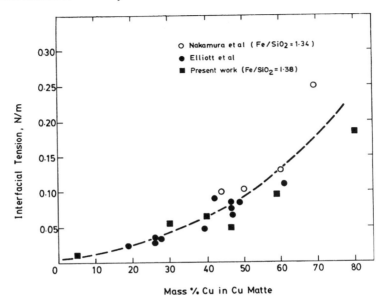

**Fig 14.** *Interfacial tension of the Cu–Fe matte/slag system at 1473 K.*

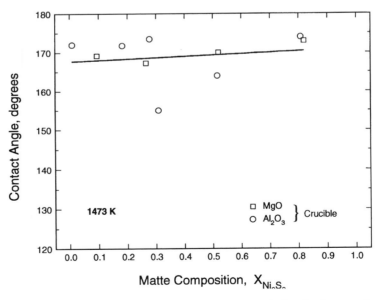

**Fig 15.** *Contact angle between Ni–Fe matte and alumina/magnesia in the presence of slag at 1473 K.*

Fig 15 shows that the contact angle of nickel mattes on alumina and magnesia in the presence of a slag phase is essentially independent of matte composition. The large contact angle (~170°on average) between the mattes and the crucible materials clearly indicates that alumina and magnesia are wetted preferentially by liquid slag.

*Conclusions*

For the pure sulphides, the surface tension is a linear function of temperature. For the pseudo-binaries, the surface tension behaved more or less with composition except for the FeS–Cu$_2$S system. The interfacial tension data show very similar behaviour for both the Cu–Fe and Ni–Fe matte/slag systems. In both cases, the interfacial tension decreases non-linearly with the addition of FeS to the matte. The interfacial tension approaches zero at pure FeS composition.

The contact angles between Ni–Fe–S mattes and various oxides (alumina, magnesia, quartz, sapphire and a high chromium oxide brick, C1215) revealed that only quartz has consistently high contact angle values (≥ 90°) over the full range of matte composition studied. The results also showed that regardless of substrate materials, the contact angle increases with the nickel sulphide content in the matte. Contact angle measurements from interfacial experiments show that the presence of a slag phase greatly increases the contact angle between the matte and the solid oxides (Al$_2$O$_3$ and MgO in this case).

## Mechanical Entrainment of Matte in Slag by Gas Bubbles

*Background*

Minto and Davenport[16] have analysed the influence of surface and interfacial tension on the entrainment behaviour of liquid droplets in a continuous liquid phase. They found that in order for the entrained droplets to be floated by gas bubbles (Fig 16b), the flotation coefficent for the matte-slag system, as defined by Eq. (1) must

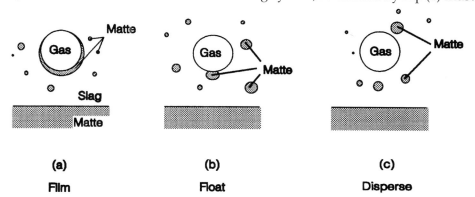

**Fig 16.** *Interaction between gas bubble and liquid droplets in a continuous phase upon contact[16]: (a) droplet forms a film on the gas bubble; (b) droplet attaches to gas bubble; and (c) no attachment.*

be greater than zero. Together with the definition of the filming coefficient, Eq. (2)[17] entrainment behaviour of droplets can be predicted from surface and interfacial tension values. For film formation to occur as illustrated in Fig 16a, the filming coefficient must be greater than zero. From Eq. (2), this implies that the surface tension of the slag must be much larger than the surface tension of the matte and interfacial tension of the system. As a result, it is more favourable for the matte droplet to form interfaces with both the slag and the gas phase. When drop flotation occurs, the flotation coefficient is greater than zero but the filming coefficient is less than zero. Under such conditions, the surface free energy of the system is readily minimised by the formation of a matte–gas interface, Fig 16c represents the scenario when the flotation coefficient is less that zero, in which case collision between matte droplets and rising gas bubble does not result in the attachment of droplets to the gas bubble,

$$\Delta \quad = \quad \gamma_{s/g} - \gamma_{m/g} + \gamma_{m/s} \qquad (1)$$

$$\phi \quad = \quad \gamma_{s/g} - \gamma_{m/g} - \gamma_{m/s} \qquad (2)$$

where:

$\Delta$ = flotation coefficient, $\phi$ = filming coefficient, $\gamma_{s/g}$ = slag surface tension $\gamma_{m/g}$ = matte surface tension and $\gamma_{m/s}$ = matte-slag interfacial tension.

Figure 17 corresponds to the calculated flotation and filming coefficients for the Cu–Fe matte in slag system and Fig 18 represents the results for the Ni–Fe matte/

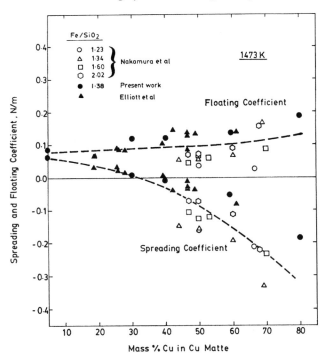

Fig 17. *Filming and flotation coefficients of the Cu–Fe matte-slag-gas system at 1473 K.*

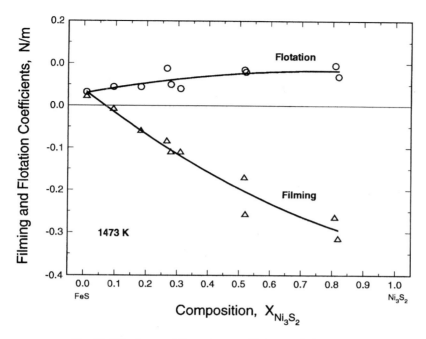

**Fig 18.** *Filming and flotation coefficients of the Ni–Fe matte-slag-gas system at 1473 K.*

slag system. As shown in both figures, the flotation coefficient remained positive for the entire composition range. Therefore, flotation of matte by gas bubbles in the slag is unavoidable regardless of matte composition. On the other hand, filming of matte on gas bubble only occurs at lower matte grades, less than ~7.5 mol% $Ni_3S_2$ for the Ni–Fe matte and less than 32 mass% Cu for the Cu–Fe matte The predicted flotation behaviour for the Ni–Fe matte slag system is very similar to that obtained for the Cu–Fe matte/slag system. From Figs 17 and 18, it can be concluded that it is impossible to avoid entrainment loss due to bubble attachment in nickel and copper matte processing.

## Refractory Infiltration

*Background*

In pyrometallurgical processes, infiltration and corrosion of refractory bricks are of great importance. Infiltration not only represents pay metal loss but it can also affect both the thermal and mechanical properties of the bricks. With the presence of infiltrated matte, the thermal conductivity of the brick becomes higher. As a result, matte penetration may advance and eventually results in a break out. Altering the mechanical properties of the brick results in early brick failure and more frequent repairs. Furthermore, due to the difference in the thermal expansion coefficient between the brick and the infiltrated material, thermal fluctuation in the furnace will cause the infiltrated bricks to spall.

The fundamental equation which governs the height of infiltration of liquid into a capillary immersed in a liquid is given by

$$h = \frac{2\gamma\cos\theta}{r\rho g} \tag{3}$$

where:

h = height of infiltration, $\gamma$ = surface tension of liquid, $\theta$ = contact angle between liquid and solid, $r$ = radius of capillary, $\rho$ = density of liquid and, $g$ = acceleration due to gravity.

This equation shows that in order to minimise infiltration the surface tension of the liquid should be small. On the other hand, the contact angle, capillary size and density of the liquid should be large. In case of refractory materials which are generally submerged in the liquid phase, the shape of the meniscus and the balance between the hydrostatic pressure and surface tension is slightly different than that given in Eq. (3) The equation for liquid penetrating a submerged porous object is given by

$$h = \frac{-2\gamma\cos(180 - \theta)}{r\rho g} \tag{4}$$

Equation (4) indicates that to minimise liquid infiltration, the density of the liquid and pore size in the submerged object must be small. Conversely, the surface tension and the contact angle between the liquid and solid should be large.

*Discussion*

Using Eq. (4), the infiltration depth of matte on various oxide materials are calculated using the data presented in Figs 8 and 11. The pores on the solid are assumed to have an average diameter of 1 mm. As indicated in Fig 19, the infiltration depth decreases with increasing matte grade for all the oxides substrates investigated. The positive depth values for C1215 (high $Cr_2O_3$ brick) indicate that penetration by nickel matte cannot be avoided by altering the composition of the matte phase. Except for quartz, all the other oxides cross the zero infiltration point as the matte grade changes. For magnesia, this corresponds to a composition of about 45% $Ni_3S_2$. From these results, it is clear that refractories containing high

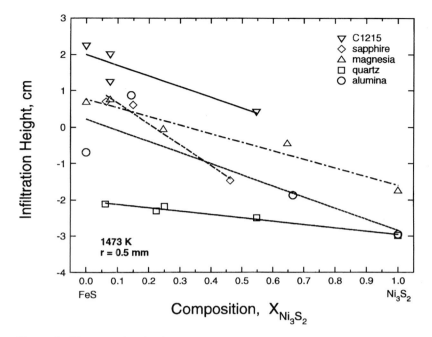

**Fig 19.** *Infiltration depth of Ni–Fe matte in various oxide materials at 1473 K. Values calculated using a pore diameter of 1.0 mm.*

$Cr_2O_3$ content such as the C1215 (94.2% $Cr_2O_3$, 3.8% $TiO_2$ typical) brick should not be used for matte containment because of its infiltration property. The other oxides with the exception of quartz are not very suitable components as refractory materials since their infiltration behaviour falls in the border line range. Quartz is the only material tested which provides good infiltration resistance. Unfortunately, quartz will flux with the iron oxides present in the system and thus cannot be used effectively as a refractory material.

When a slag phase is present together with the matte, the wetting behaviour of the matte on the solid oxide surfaces changes dramatically. This is demonstrated by the contact angle measurements obtained from interfacial tension experiments. Since the penetration depth of the matte into the pore is a direct function of the angle, the presence of the slag phase improves the penetration resistance of the refractories. This effect is illustrated in Fig 20. Comparing the penetrating depth of matte in magnesia, infiltration is greatly suppressed in the presence of a slag phase as shown by the large negative infiltration depth values. This observation suggests that for each new campaign it is beneficial to melt slag in the furnace prior to the addition of matte into the vessel. In this case, the new refractories will be covered by a layer of slag which prevents the matte from contacting the refractory surface. The presence of slag layer should reduce the amount of matte penetration into the refractory.

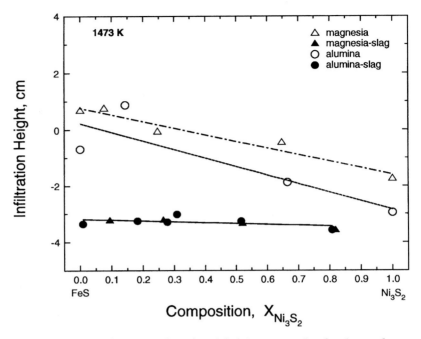

**Fig 20.** *Calculated infiltration data for nickel-iron matte in alumina and magnesia as affected by the presence of a slag phase at 1473 K.*

Although much has been said on infiltration, it is important to keep in mind that in all commercial operations the height of both the matte and slag phase in the furnace generally fall in the meter range. Therefore, due to this large hydrostatic head, some infiltration of matte and slag into the refractory is inevitable.

## Slag Cleaning by Electrocapillarity

*Background*

The most common practice of slag cleaning is by holding the slag at elevated temperature in an electric furnace to allow the heavier matte or metal droplets to settle under gravity. However, it is believed that it may be possible to take advantage of the electric field present in the electric furnace to further enhance the settling process. The improvement in settling is achieved through a phenomenon known as electrocapillarity.

Electrocapillarity was first proposed by Christiansen[18] in 1903. To understand the electrocapillary phenomenon better, consider a liquid droplet which is ideally polarised in a solution. A double layer is formed and the resultant charge distribution at the interface can be assumed to be random as shown in Fig 21a. When an external

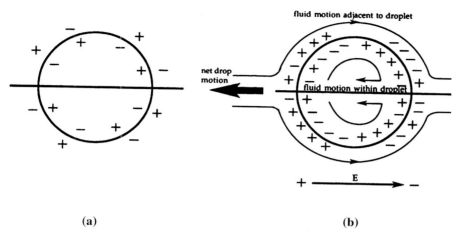

**Fig 21.** *Charge distribution of a droplet in solution (a) in the absence of an electric field and (b) under the influence of an electric field, E. The charges are randomly distributed in (a) while the charges are polarised in (b) as a result of the external field.*

electric field is applied, there will be a redistribution of charges such that there are more positive charges on one side of the droplet and fewer on the other (Fig 21b). The imbalance in charges sets up an interfacial tension gradient along the interface. This results in a flow of liquid within the drop from a region of low interfacial tension to a region of high interfacial tension along the interface. This surface flow in the drop will also drag adjacent layers of liquid in the solution. As a result, the liquid drop will move in the solution against the direction of the surface flow.

Levich[19] has developed from first principles the expression for describing the electrocapillary migration velocity, $U$, of an ideally polarised droplet:

$$U = \frac{\varepsilon E a}{2\mu_m + 3\mu_d + \varepsilon^2/\chi} \tag{5}$$

where $\varepsilon$ is the surface excess charge density ($Cm^{-2}$), $E$ is the applied potential field ($Vm^{-1}$), $a$ is the droplet radius ($m$), $\mu_m$ is the viscosity of the melt, $\mu_d$ is the viscosity of the droplet and $\chi$ is the electrical conductivity of the melt ($\Omega^{-1}m^{-1}$). Except for $\varepsilon$, the rest of the variables are either physical properties or experimental parameters. $\varepsilon$ can be calculated via the Lippmann equation:

$$\frac{\partial\gamma}{\partial\varphi} = -\varepsilon \tag{6}$$

where $\gamma$ is the interfacial tension and $\varphi$ is the potential drop across the droplet/melt interface. This is the slope of the electrocapillary curve as shown in Fig 22b. The electrocapillary motion of the droplet (Fig 22a) is superimposed above the

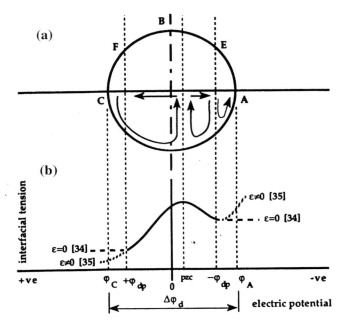

Fig 22. (a) Electrocapillary motion and (b) electrocapillary diagram when a droplet is partially polarised.

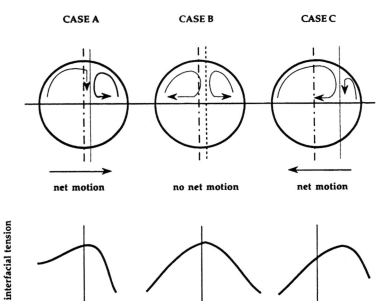

Fig 23. (a) possible electrocapillary motion due to (b) electrocapillary curves with a single maxima. Case A, net right motion; Case B no motion; Case C, net left motion.

electrocapillary diagram. An extension of the above analysis is shown in Fig 23 where the motion of the droplet is governed by the shape of the electrocapillary curve, Thus, by determining the interfacial tension between the metal or matte/melt as a function of electric potential, it is possible to predict both the direction and speed of migration of these droplets.

A large number of research studies have been carried out in the former USSR on the subject of electrocapillary motion of droplets in metallurgical slags. On the other hand, works of this nature were very limited in the Western literature. Therefore, in addition to providing a better fundamental understanding of this subject, it is hoped that this study will also stimulate interest in this exciting mode of droplet migration control in metallurgical systems.

## Results and Discussion

*Constant Matte and Slag Composition*
Figure 24 shows the migration rate of various droplets in synthetic fayalite at 1523 K while Fig 25 shows the migration rate of $Cu_2S$ in similar slag under similar conditions as a function of electric potential.[20] Positive velocity on the graphs indicates migration to the anode. Eq. (5) can be qualitatively verified from the cases studied in Figs 24 and 25. The migration rate increases as the droplet size or applied potential is increased.

Figure 24 is particularly interesting as it shows that the droplet can migrate to

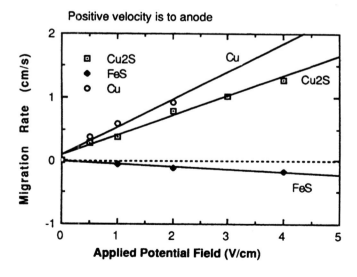

Fig 24. *Migration rates of copper, copper sulphide, and iron sulphide droplets on the surface of synthetic fayalite slag (70% FeO) under and electric field at 1523 K.*

either electrode. When a Cu₂S droplet was placed near the cathode and an FeS droplet near the anode, it was observed that the droplets migrated to the opposite electrodes when the electric field was applied. In fact, the droplets migrated past each other across the slag surface. Such independent motions suggest the possibility of specific control of the droplet's migration behaviour which can be advantageous during the slag cleaning operation.

Figure 25 shows a parabolic-type relationship rather than the linear function as

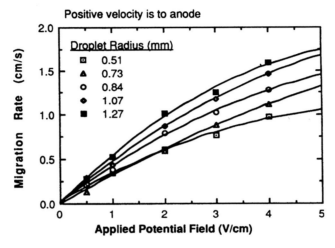

**Fig 25.** *Migration rates of copper sulphide droplets on the surface of synthetic fayalite slag (70% FeO) as a function of applied potential at 1523 K.*

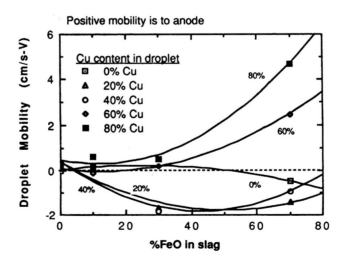

**Fig 26.** *Mobility (U/Ea) of Cu–Fe matte as a function of FeO content in slag at 1573 K.*

expressed in Eq (5). This observation can be attributed to partial polarisation:

$$U_{pp} = \frac{\varepsilon E a}{2\mu_m \left(1 + \dfrac{a}{2\chi\omega}\right)} \qquad (7)$$

where $U_{pp}$ is the terminal velocity under partial polarisation and $\omega$ the interfacial resistance, defined by:

$$\omega = \frac{RT}{zFi_{lim}} \qquad (8)$$

is a measure of the partial polarisation at the interface. Partial polarisation occurs when there is ion exchange or charge transfer across the interface. Thus $i_{lim}$, the limiting current density, is non zero and $U_{pp}$ is depressed with respect to $U$. Levich explained that this decrease in droplet velocity is caused by the resulting conductance at the interfacial region where the electromotive force of a partially polarised droplet is weaker than that of an ideally polarised droplet.

*Variable matte and Slag Composition*
In this set of experiments, both the initial matte and slag compositions were varied. The matte composition ranges from pure FeS to pure $Cu_2S$ while the slag composition ranged from fayalite (70% FeO, 30% $SiO_2$) to a calcium–aluminum silicate (40% CaO, 20% $Al_2O_3$, 40% $SiO_2$). The experiments were carried out at 1573 K.

Fig 26 shows the effect of FeO content on droplet mobility ($U/Ea$). It is clear from the displayed data that the velocity (thus both the speed and direction) of the droplet is dependent on the composition of both the matte and slag.

If $\varphi_A$ or $\varphi_C$ (Fig 22a) is such that it exceeds the decomposition potential, $+\varphi_{dp}$ or $-\varphi_{dp}$ of the chemical species in the droplet, then redox reactions will occur at the interface and the droplet is no longer ideally polarised. With ion and electron exchanges occuring at the interface, the charge density and distribution will be different. As a result, the interfacial tension gradient along the interface will also be altered. Ultimately, the migration rate of the droplet will be affected. Following the same line of reasoning, any kind of reaction, either chemical or electrochemical, which could alter the interfacial tension gradient along the interface will have an effect in the mobility of the droplet.

*Electrocapillary Diagram of the $Cu_2S$–FeS/Fayalite Slag System*
Two sets of preliminary determination of the electrocapillary diagram of the $Cu_2S$–FeS/fayalite slag system are plotted in Fig 27. There is substantial scatter in the data; however, it is still clear that there is a gradual shift up in the shape of the electrocapillary curves. The vertical shift in the curves accounts for the higher

velocity due to the higher interfacial tension gradient. Note that the increase in interfacial tension is not linear with composition.

There appears to be qualitative agreement between the data in Fig 26 and the

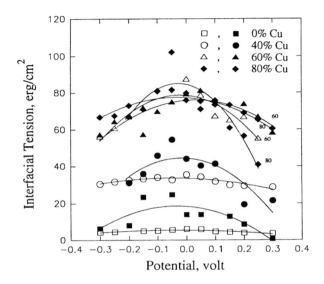

**Fig 27.** *Electrocapillary diagram of the Cu$_2$S–FeS matte/fayalite slag system at 1523 K.*

electrocapillary curves in Fig 27. At pure FeS (0% Cu) composition, the mobility of the droplet is low. This corresponds with the flat slope of the electrocapillary curve in Fig 27. The droplets containing 60 and 80% Cu show high mobility (Fig 26), an observation which corresponds to a steep slope on the electrocapillary curves as shown in Fig 27. Furthermore the maxima of the electrocapillary curves tend to hover around zero potential. This may help explain why the droplet can migrate to the anode and cathode, for it indicates that a proper balance of the forces within the flow loops inside the droplet is the controlling factor governing droplet migration direction. If the curves would have been predominantly on one side of the diagram, one would expect the droplet to have only a specific migration direction under all slag and drop composition. As it is, Fig 27 is in qualitative agreement with the magnitude of the measured electrocapillary speeds.

Electrocapillarity, as applied to slags, offers an exciting possibility for controlling the migration direction and speed of second phase metal and matte droplets. It requires only a direct current potential field and the passage of current is not necessary, thus minimising energy consumption. The speed generated by electrocapillary forces is of the order of cm/s and thus offers a much higher rate of slag cleaning when compared to traditional gravimetric settling.

## References

1. S.W., Ip, and J.M. Toguri *Met. Trans. B*, 1992, **23B**, 303–311.
2. Y. Rotenberg, L. Boruvka, and A.W. Neumann. *J. Colloid Inter. Sci.*, 1983, **93**, 169–183.
3. S.W. Ip. PhD Thesis, University of Toronto, Canada, 1992.
4. S.J. Sugden, *Chem. Soc.*, 1922, **121**, 858–66.
5. S.J. Sugden, *Chem. Soc.*, 1924, **125**, 27–31.
6. M. Kucharski, S.W. Ip, and J.M. Toguri, accepted for publication in *Can. Met. Quart.*, Oct. 1993
7. T. Fujisawa, T. Utigard, and J.M. Toguri, *Can. J. Chem.*, 1985, **63**, 1132–1138.
8. G. Liu, J.M. Toguri and N.M. Stubina, *Can. J. Chem.*, 1987, **65**, 2779–2782.
9. S. Tokumoto, A. Kasama and Y. Fujioka. *Technol. Report Osaka Univ.*, 1972, **22**, 453–463.
10. A.V. Vanyukov and V. Ya. Zaitsev, *Slags and Mattes in Non-Ferrous Metallurgy*, Moscow, 1969, p95.
11. S.E. Vasiburd, A.G. Ryabko, Yu.V. Fisher, A.V. Taberko, I.N. Zedina and M.M. Marshak: *Melts*, 1988, **1**, 28–36.
12. Q. Liu, K. Huang, D. Ye and X. Chen. *J. Central-South Inst. Min. and Metall.* 1986, No. 3, 87–94.
13. V. Ya. Zaytsev, A.V. Vanyukov and V.S. Kolosova. *Russ. Met.*, 1968, No. 5, 29–33.
14. J.F. Elliott and M. Mounier. *Can. Metall. Quart.*, 1982, **21**, 415–428.
15. T. Nakamura, F. Noguchi, Y. Ueda and S. Nakajyo. *J. M.M.I.J.*, 1988, **104**, 531–36.
16. R. Minto and W.G. Davenport. *Trans. Inst. Min. Metall.*, 1973, **82**, C59-62.
17. W.D. Harkins. *The Physical Chemistry of Surface Films*, Rheinhold, New York, 1952.
18. S. Christiansen. *Ann. Phys.*, 1903, **12**(4), 1072.
19. V.G. Levich. *Physiochemical Hydrodynamics*, Prentice-Hall, Englewood Cliffs, 1962, pp. 472–531.
20. R.T.C. Choo and J.M. Toguri. *Can. metall. Quart.*, 1992, **31**, 113-126.

# Physical Property Measurements for Melts at High Temperatures

KENNETH C. MILLS,

*Division of Materials Metrology, National Physical Laboratory, Teddington, Middlesex TW11 0LW, UK*

## Abstract

Physical properties of melts at high temperatures are greatly affected by reactions between containers and the melt. Since these reactions can lead to erroneous results, in recent years contactless methods have been developed to measure thermodynamic properties, surface tensions, densities, enthalpies, heat capacities and emissivities. Most of these measurements were carried out using electromagnetic levitation and were thus confined to metals. However, acoustic, aerodynamic and electrostatic levitation has been successfully developed over the last decade and should soon lead to measurements on non-metallic melts. Experiments carried out in microgravity have provided values of the diffusivity and thermal conductivity of melts which are free from the effects of convection.

## Introduction

Over the last decade mathematical modelling has become an established tool to improve process control and efficiency and product quality for a wide variety of high temperature processes, e.g.. Casting processes, steelmaking, coal gasification etc. In many cases these models have been developed to the stage where the prime requirement is for accurate property data for the materials involved in the process. Accurate values are needed for both thermodynamic properties and those physical properties involved in the heat and fluid flow in the process such as thermal conductivity, emissivity, heat capacity, enthalpy, density, viscosity, surface tension. The paucity of reliable data for these properties at high temperature can be traced directly to the experimental difficulties encountered in making these measurements. The unofficial First Law of High Temperatures states that *At high temperature everything reacts with everything else* and the Second Law that *They react very quickly and it gets worse as the temperature increases.* Reactions between melts and containers can have a marked effect on property measurements particularly where liquids and high temperatures involved. Consequently, in order to minimise these reactions it is important to use non-contact methods such as levitation[1] or the direct pulse technique in which wire samples are rapidly heated (over a period of milliseconds) to high temperatures.[2] Although property data for liquids have been obtained with the latter method, this technique is designed primarily for solid-state measurements. Consequently most physical property data for reactive liquids at high temperatures have been obtained with levitation techniques.

There are several different techniques which can be used for levitation as follows:

*Electromagnetic levitation*

Levitation is achieved by coupling a special coil to a RF power source. The repulsive force required for levitation is provided by the reaction of the eddy currents (induced in the surface of the sample) to the primary field produced by the lower windings of the coil. The upper counterwindings create a null point in the magnetic field which permits stable positioning of the sample. 'Cold crucibles' provide and alternative form of electromagnetic levitation in which levitation results from the repulsion of like charges on the sample and crucible. Both forms of electromagnetic levitation are restricted to materials with high electrical conductivity, e.g. metals and some carbides and sulphides.

*Acoustic levitation*

High intensity sound waves are used to generate sufficient force to counteract the gravitational forces.[3] A reflector is used to produce a minimum acoustic pressure zone and thus provide stable positioning of the sample. This form of levitation can be applied to both conducting and non-conducting materials but cannot be used in vacuum.

*Aerodynamic (gas jet) levitation*

High velocity gas flows from specially-designed nozzles are used to levitate the sample. The spreading action of the jet can be used to position the sample which can be further improved by electromagnetic stabilisation. Liquids tend to fragment in the gas flow but recently, investigations on molten metals have been successfully carried out.[4,5]

*Electrostatic levitation*

The charge-carrying sample is levitated in an electrostatic field generated between two or more electrodes. Since there is no potential minimum for an electrostatic field, stable positioning of the specimen can only be obtained with a feedback mechanism.[6]

*Microgravity*

Drop tubes, parabolic and space flights have all been used extensively to provide levitation under reduced gravity conditions. Since microgravity experiments are expensive and are very time-consuming it is important that they should be devoted to those physical property measurements which can not be obtained in any other way, viz (i) measurements which are affected by convection (thermal conductivity and diffusivity) and (ii) for highly undercooled melts.

In this paper, the levitation and microgravity techniques developed for physical property determination are described, and conventional methods are outlined where no satisfactory containerless method exists. This paper is based on the author's expertise and should not be regarded as an exhaustive review such as that recently published by Margrave and colleagues.

## Levitation Measurements

As we have seen, several contactless methods are available for physical property measurements, however, the author is not aware of any accurate physical property measurements which have been carried out using 'cold crucible'* levitation. The use of acoustic, aerodynamic and electrostatic levitation is relatively new and their application to high temperature systems has not been fully developed yet. In contrast, the technology of electromagnetic levitation is relatively mature, thus most of the measurements described below have been carried out using electromagnetic levitation and consequently apply primarily to metallic systems. Hopefully, the development of acoustic, aerodynamic and electrostatic levitation techniques will progress in the next decade to the stage where these measurements can be obtained for non-conducting samples.

## Thermodynamic Properties

Measurements of the chemical activities of binary alloys were derived[9] using electomagnetic levitation (Fig 1a) by levitating an alloy of known composition at a constant temperature in an inert atmosphere and condensing the vapour on the

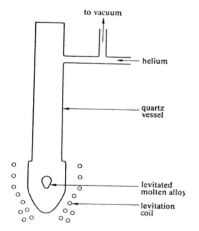
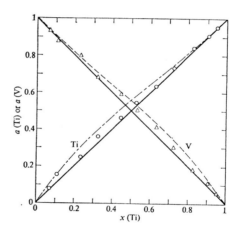

**Fig. 1.** *(a) Vaporisation cell; (b) Activities a(Ti) and a(V) in Ti–V system as a function of mole fraction Ti (x(Ti). Measurements for 2073 K with He atmosphere: O, a(Ti); Δ, a(V); – –, – –, values reported for solid phase.*

---

* Kawase *et al*[8] deduced surface tension values for metals from the shape of the meniscus of melts held in cold crucibles, but this can not be considered as an accurate method for determining surface tension.

sides of the silica tube. The composition of the vapour was determined by chemical analysis. Measurements were carried out at temperatures up to 2300°C for reactive systems (Fig 1b). Activities (a) for the Ti–V system[10] were derived by using Eq. (1).

$$\ln a(Ti) = \frac{D_{Ti}}{D_v} \int_{x(Ti,g) = 1}^{x(Ti,g) = x(Ti,g)} \frac{x(V,1)}{x(V,g)x(Ti,g)} \, dx(Ti,g) \qquad (1)$$

where $D$ is the interdiffusivity of metal and inert gas, $x$ the mole fraction, and $l$ and $g$ respectively, represent the liquid (i.e. the droplet) and the gas (i.e. the condensed vapour). The activity of vanadium was derived by Gibbs–Duham integration.

Betz and Frohberg[11] have determined (i) the enthalpy of solution of binary alloys, such as the Mo–Si system by adding solid Si to a molten Mo droplet and monitoring the temperature effect and (ii) the solubility of oxygen and nitrogen in liquid metals such as Mo.[12]

The thermodynamics and kinetics of slag/metal reactions at high temperatures have been investigated using levitation methods by Jahanshahi and Jeffes[13] and Creehan and Grieveson[14] at Imperial College. The slag was added by either bringing the metal droplet in contact with slag held on a retractable alumina table[13] or by inserting slag into a hole drilled in the specimen and sealing with a plug of the sample.[14] This method has been applied to the measurement of the kinetics of dephosphorisation of ferrous melts.[14]

*Vapour Pressure*

Levitation melting has been used to determine the vapour pressure of liquid iron using the carrier gas transpiration method;[15] the results obtained were found to be in good agreement with accepted values. Enthalpies of vaporisation were derived for levitated samples of Mo, W and Zr by measuring the laser induced fluorescence as a function of temperature.[16]

Vaporisation rates have been derived at the same time as chemical activity measurements by monitoring the mass loss of the sample when held at constant temperature for a known time.[17]

*Surface Tension Measurements*

The surface tensions of liquid metals are very dependent upon the concentrations of surface-active elements (such as oxygen or sulphur) present. For example, the addition of 50 ppm sulphur to iron causes (i) a decrease of 30% in the surface tension ($\gamma$) and (ii) the temperature coefficient ($d\gamma/dT$) to change from negative to positive. Thus reactions between liquid metals and oxide ceramics at high temperatures lead to oxygen pick-up by the melt and to low surface tension values. In order to eliminate this source of contamination, Fraser *et al.*[18] levitated metal samples and derived the surface tension by monitoring the frequency of the natural oscillation

($\omega_R$)of the droplet using the Rayleigh relationship shown in Eq. (2), where $m$ is the mass of the drop:

$$\gamma = 3 \pi m \omega_R^2 / 8 \qquad\qquad (2)$$

This method was further developed, in our laboratory, by introducing a Fourier waveform analyser for accurate determination of the oscillation frequency.[19] The apparatus is shown in Fig. 2. It was found that a frequency spectrum containing 3 or 5 peaks (Fig. 3) was obtained and not the single frequency predicted by theory.

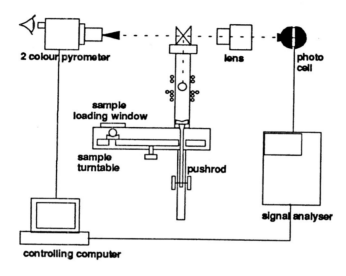

**Fig. 2.** *Oscillating drop method for surface tension measurements.*

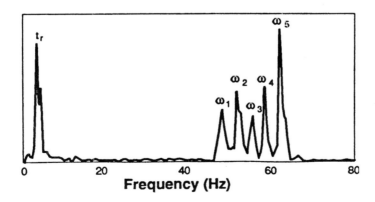

**Fig. 3.** *Typical frequency spectrum for a levitated drop.*

Recent work[20] has established that the oscillations arise from both translational movements and the effect of magnetic pressure from the coil which results in an asymmetrical drop. Cummings and Blackburn[20] have shown that the Rayleigh frequency, $\omega_R$, can be derived from 5-peak spectra by use of Eq. (3), where the subscripts 1–5 refer to the various peaks, tr to the translational frequency, $g$ is the gravitational constant and a the radius of the drop

$$\bar{\omega}_R^2 = 1/5 \ \Sigma \ \omega_1^2 + \omega_2^2 + \omega_3^2 + \omega_4^2 + \omega_5^2 - \omega_{tr}^2 \left( 1.9 + \frac{0.6g}{a(2\pi\omega_{\kappa \, tr})^2} \right)^2 \qquad (3)$$

The translational frequency, $\omega_{tr}$, is dependent upon mass and the specific levitation facility.[21] Consequently collaborative work was carried out between our laboratory and DLR in Cologne. Experiments carried out on nickel (Fig. 4), iron and gold indicated that the surface tension results derived using Eq. (3) showed no mass

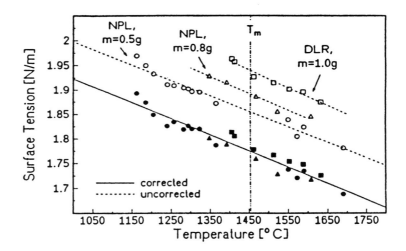

**Fig. 4.** *Surface tension of Ni as a function of temperature; – –, open symbols, values derived by Eq. 2 using the mean frequency; —, closed symbols, values derived with Eq. 3.*

dependence, and the values obtained by the two laboratories were in excellent agreement.[21] It can be seen from Fig. 4 that the surface tensions for the undercooled melt could be derived by extrapolation of values for the liquid state.

The levitated drop method can also be applied to semi-conductor systems such as silicon. The sample was placed on a retractable, boron nitride table and heated inductively in the coil until it attained a temperature where its conductivity was sufficient to provide levitation.[22] However, problems were encountered with excessive translation motion in the vertical direction due to differences in 'coupling' in the electromagnetic field. This problem could probably be overcome by heating the sample resistively.

Measurements of surface tension of non-conducting liquids have been obtained by ultrasonic levitation but only at room temperature.[23] In principle, this method could be applied to high melting, non-metallic systems but will require some development.

## Density Measurements

Density measurements were first obtained with the levitated drop method by Ward co-workers.[24] The sample of known mass is levitated and photographed simultaneously in the vertical and horizontal directions and the volume of the drop determined (see Fig. 5). It is important that the drop should be near spherical; for

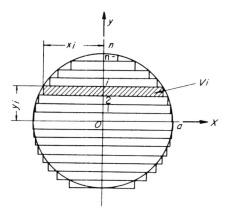

**Fig. 5.** *Determination of the volume of a levitated drop.*

Fe, Ni and Co with surface tensions these conditions are satisfied with sample masses of 0.4–0,5 g but low surface tension systems (e.g. Sn) would require much smaller masses and thus reduce the accuracy of the method. Preliminary work in our laboratory indicated that densities with an accuracy of ± 2% can be derived with this method by taking the mean of several determinations. This can be compared with uncertainties of around ± 1-2% for the Archimedian method, but the accuracy of the levitated drop method can probably be improved and it does have the advantage that it can be applied to density measurements covering a wide temperature range.

Ronchi[25] has measured thermal expansion coefficients of liquid $Al_2O_3$ by levitating particles using ultrasonic levitation and then heating the sample with a laser beam. The principal difficulties encountered were associated with the instability of the sample and with temperature equilibration.

## Enthalpy, Heat Capacity Measurements

Drop calorimentry is an accepted method for measuring the enthalpies of materials at high temperatures. However, it is difficult to obtain accurate enthalpy measurements for the molten range since ceramic containers have a relatively high heat capacity when compared with those for metals (e.g. Pt, which is used for the solid range) and this leads to large experimental errors. Furthermore, ceramic crucibles are vulnerable to thermal and mechanical shock which could result in molten metal escaping from the crucible and damage to the calorimeter. In order to overcome these problems, levitated drop calormeters have been developed.[26–28] These have the further advantage that they can be applied to large temperature ranges (temperatures > 3000°C has been achieved by using a laser to provide additional heating). A schematic diagram of the levitated drop calorimeter developed in our laboratory is shown in Fig. 6 and it can be seen that the temperature

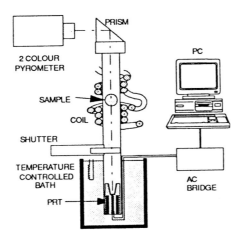

**Fig. 6.** *Schematic diagram of levitated drop calorimeter.*

rise (typically 4°C) of the copper receiving block can be measured to better that $10^{-3}$ °C. Experimental uncertainties in enthalpy measurements are typically ± 1%; the principal source of uncertainty lies in the temperature measurement provided by the two colour pyrometer.

Heat capacities ($Cp$) are derived by differentiation of enthalpy-temperature relationships. Although solid-state transitions can lead to an increase in the uncertainty of the $Cp$ values, those derived for the liquid state should be reasonably accurate providing there are a sufficient number of measurements and the temperature readings are accurate. Hiernaut and Ronchi[29] determined heat capacities of W and $Al_2O_3$ by analysis of the heat losses of acoustically levitated spheres using laser beams to treat the specimen; uncertainties in $Cp$ were ± 10-15%.

*Emissivity*

Measurements of the spectral emissivities of molten levitated drops have been made by Hansen *et al*[30] using electromagnetic levitation and rotating-analyser ellipsometry to determine the intensities of light reflected from the sample at various angles;

values of the normal emissivity and reflectivity were derived from these measurements.

## Measurements in Microgravity

### Thermal Conductivity

It is difficult to eradicate convection from thermal conduction measurements even at ambient temperatures and convective contributions lead to high values of the conductivity. Convection can be eliminated by working in microgravity. Nakamura et al.[31] measured the thermal conductivities of mercury and molten InSb under terrestial and microgravity conditions (i.e. in a drop tube and a parabolic flight) using the line source method. They were able to show that significant convectional contributions were obtained for experiment durations which exceeded 1 second (Fig. 7). The probe was coated in ceramic to eradicate electrical leakage to the melt.

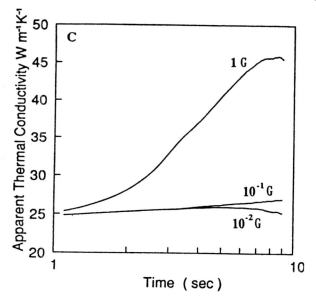

**Fig. 7.** *Thermal conductivity of InSb melts derived for terrestial and microgravity conditions.*

### Diffusivity

It has been recognised that convection may assist the diffusion process in liquids and may lead to enhanced values for the measured diffusion coefficient. Frohberg et al.[32] measured the self diffusion coefficients of $Sn^{112}$ in $Sn^{119}$ in microgravity by the radioactive tracer method and found that the diffusion coefficients in microgravity were appreciably lower than those recorded on earth (Fig. 8).

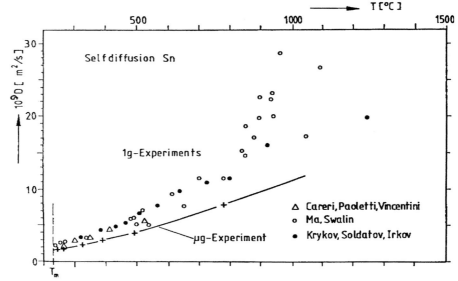

**Fig. 8.** *Comparison of self diffusion coefficients of Sn melts obtained under microgravity and terrestial conditions.*

## Electrical Conductivity

Electrical conductivities for the liquid state are not affected by convection, thus the determination of this property in microgravity arises from an interest in the undercooled state. Measurements of the electrical conductivity of liquid nickel is to be attempted on the TEMPUS flight using a contact-free inductive method.[33] The sample is placed inside a solenoid which is part of an oscillatory circuit (Fig. 9). Any change in the electrical conductivity of the sample alters the total impedance and hence leads to a subsequent change in the alternating current. Since the latter can measured accurately, the specific electrical conductivity can be derived for the sample.

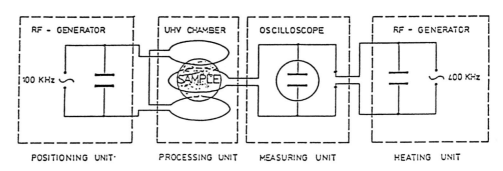

**Fig. 9.** *Schematic diagram of the inductive method for measuring electrical conductivities.*

*Heat capacity*

It was pointed out above that the effects of convection on the cooling characteristics of a molten, acoustically-levitated sphere were found to be difficult to quantify and lead to uncertainties of $> \pm$ 10% in Cp values.[29] Consequently, microgravity is an attractive method to obtain direct measurements of heat capacity of liquid metals; measurements will be carried out on the TEMPUS flight using the non-contact modulation calorimentry[34] and values will be obtained for the supercooled state. In this method the sample is held in an ultra-high vacuum furnace and is heated in an RF heating coil for which the heating power is modulated by superimposing a sinusoidal signal to the RF field. The temperature of the sample is monitored continuously with an InSb pyrometer capable of measuring to $\pm$ 1°C and heat capacities can be derived after inserting a value for the emissivity of the metal. The apparatus was tested on earth using a Nb sphere suspended by a fine wire and the measured values of Cp were within $\pm$ 1% of recommended values.

*Viscosity*

Measurements of the viscosities reported for liquid metals are very disparate as can be seen for the values for molten Fe[35] shown in Fig. 10. Egry and co-workers[36] have proposed that viscosity measurements can be derived from the damping characteristics for the oscillations of a molten, levitated sphere. However, oscillations could arise from several sources, e.g. magnetic effects and natural oscillations. The proposed method is only valid when there is a dominant source for the oscillations.

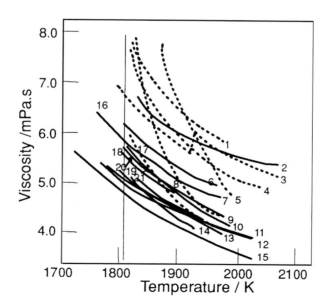

**Fig. 10.** *Viscosity-temperature relationships reported by various investigators for molten Fe.*[35]

Thus it is proposed to carry out experiments in microgravity in which external, ultrasonic oscillations will be imposed on the sample.

## Conventional Methods

Some physical properties are sensitive to small changes in composition (e.g. surface tension) whereas others (e.g. density) may not be sensitive to these changes. Consequently, conventional techniques can provide accurate values for some property measurements but errors will increase when the measurements involve reactive metals (e.g. Ti alloys or superalloys) or when using very high temperatures. Thus it is important to minimise sample/container reactions by careful selection of container materials; these are outlined on Table 1.

Table 1. Container materials used for high temperature measurements

| SAMPLE | ATMOSPHERE | CONTAINER |
|--------|------------|-----------|
| Metals | Reducing, neutral | $Al_2O_3$, MgO, $ZrO_2$, $ThO_2$, BN, C |
| Oxides | Reducing, neutral | Mo, W, Ta, Fe, C |
| Slags | Oxidising, neutral | Pt alloys, Ir |

In our experience graphite is not a satisfactory container material for accurate measurements of the physical properties since it reacts with both slags and metals. Figure 11 shows the results of interlaboratory comparison for the viscosity of a molten $Fe^{35}$.

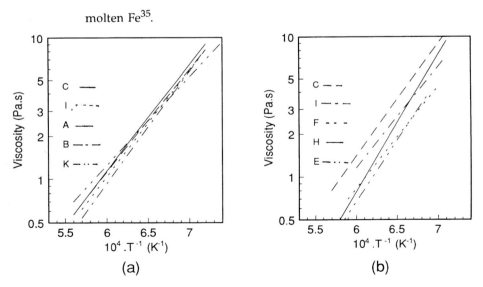

**Fig. 11.** *Viscosities of a reference $Li_2O–Al_2O_3–SiO_2$ slag as a function of temperature, using (a) Mo, Pt containers, (b) graphite containers.*

molten oxide reference material[37] and indicates that whereas there is good agreement between the results in Mo and Pt, those obtained with graphite show considerable scatter due to reaction with the molten sample. Another contributory factor may be the wetting characteristics since carbon is non-wetting cf. Mo and Pt which are wetted by the melt. Iida and Guthrie[35] have proposed that wetting characteristics of the metal on the crucible affect the viscosity measurements derived for pure metals by the oscillating crucible method. When making measurements at high temperatures with conventional techniques it is preferable (i) to use the most inert container available, (ii) to use non-contact methods wherever possible.

## Conclusions

1. Considerable progress has been made in the measurement of accurate thermophysical properties of metallic melts high temperatures by using electromagnetic levitation.

2. The recent development of acoustic, aerodynamic and electrostatic levitation should lead to thermophysical measurements for non-conducting samples.

3. Microgravity provides a method for obtaining accurate values for the thermal conductivities and diffusivities of melts free from convectional effects.

## Acknowledgements

The author gratefully acknowledges the valuable discussions held with Professor I. Egry (DLR, Cologne), Professor P. Grieveson (Imperial College) and R.F. Brooks (NPL).

## References

1. T. Baykara, R.H. Hauge. M. Norem, P. Lee and J.L. Margrave: *High Temp. Sci.* (in press).
2. A Cezairliyan: 'Pulse Calorimetry' in *Compendium of Thermophysical Property Measurement Methods*, K.D. Maglic, A. Cezairliyan and V.E. Peletoky eds Plenum, New York, Ch. 16, p.p. 643-668.
3. E.H. Trinh, J. Robey, A. Arce and M. Gaspar: 'Experimental Studies in Fluid Mechanics and materials Science Using Acoustic Levitation' in *Materials Processing in the Reduced Gravity Environment of Space*, MRS Symp. Proc., 87, R.H. Doremus and P.C. Nordine, 1987, p.p. 57–69.
4. J.P. Coutures, J.C. Rillet, D. Billard and P. Coutures: 'Contact,ell Treatment of Liquids in a Large Temperature Range by an Aerodynamic Levitation Device and laser heating,' *Proc. 6th Europ. Symp. on Mater. Sci. unser Microgravity Conditions*, Bordeaux, 1986, ESA SP-256, p.p. 427-430.
5. D. A. Winborne, P.C. Nordine, D.E. Rosner and N.F. Marley: 'Aerodynamic Levitation Technique for Containerless High Temperature Studies on Liquid and

Solid Samples,' *Metall. Trans.*, 1976, **7B**, 711–713.

6. W.K. Rhim, S.K. Chung and P. Elleman: 'Electrostatic Levitators and Drop Dynamics Experiments,' *Proc. 7th Europ. Symp. on Mater. Sci. in Microgravity*, Oxford, ESA SP-295, 1989, p.p. 629-638.

7. J. Margrave: 'Thermophysical Properties of Liquid Metals at High Temperatures by Levitation Methods', *Mater. Sci. Eng.* (in press).

8. Y. Kawase, Y. Murai and N. Hayashi: 'Numerical Analysis for Molten Metal Shape Taking into Account Electromagnetic Force and Surface Tension', *Sumitomo Light Metal Technical Repts.* 1989, 30 (2), 1–8.

9. K.C. Mills, K. Kinoshita and P. Grieveson: 'A Thermodynamic Study of Liquid Fe + Ni Alloys Using Electrmagnetic Levitation' *J. Chem. Therodyn.*, 1972, **4**, 581–590.

10. K.C. Mills and K. Kinoshita: 'Activities of Liquid Ti + V Alloys', *J. Chem. Thermodyn.*, 1973, **5**, 129–133.

11. G. Betz and M.G. Frohberg: *Metall.*, 1981, **35**, 299.

12. H, Domke and N.G. Frohberg: *Z. Metallkunde*, 1965, **65**, 615.

13. S. Jahanshahi and J.H.E. Jeffes: *Trans. Inst. Min. Metall.*, 1981, C**90**, 138.

14. R.D. Creehan and P. Grievson: see R.D. Creehan: 'The Role of Slag Chemistry in Dephosphorisation: An Equilibrium and Kenetic Study', PhD Thesis, Imperial College, University of London, 1983.

15. A.G. Svyazhin, A.F. Vishkarev and V.I. Yavoyskiy: 'Determination of the Vapour Pressure of Liquid Iron Using Levitation Melting,' *Russ Metall* 1968, **5**, 47–52.

16. R.A. Schiffman and P.C. Nordine: *Proc. of MRS Symp.*, 1987, **87**, 339.

17. E.T. Turkdogan and K.C. Mills: 'The Theory of Enhancement of Diffusion–Limited Vapourisation Rates by a Convection-Condensation Process'., *Trans. Am. Inst. Min. Metall. Engrs.*, 1964, **230**, 750–753.

18. M.E. Fraser, W.H. Lu, A.E. Hamielec and R. Murarka: 'Surface Tension Measurements on Pure Liquid Iron and Nickel by an Oscillating Drop Technique., *Metall.Trans.*, 1971, **2**, 817–823.

19. B.J. Keene, K.C. Mills and R.F. Brooks: 'Surface Properites of Liquid Metals and Their Effects on Weldability,' *Mater. Sci. Technol.* 1985, **1** 568–571.

20. D. Cummings and D. Blackburn: 'Oscillations of Magnetically Levitated Aspherical Droplets'., *J. Fluid Mech.*, 1991, **224**, 395–416.

21. S. Sauerland, R.F. Brooks, I. Egry and K.C. Mills: 'Magnetic Field Effects on the Oscillation Spectrum of Levitated Drops', Proc. of TMS Annual Conf.Denver, Feb. 1993, *Containerless Processing*, edited W Hofmeister, pp 65-69.

22. R.F. Brooks: 'The Surface Tension of Fe–Si Alloys by the Levitated Drop Technique', MSc Thesis, UMIST, Manchester, 1990.

23. E. Trinh, A. Sverin and T.G. Wang: 'Experimental Study of Small Amplitude Drop Oscillation in Immiscible Liquid System', *J. Fluid Mech.*, 1982, **115**, 453–474.

24a. El-Mehairy and R.G. Ward: 'A New Technique for the Determination of Density of Liquid Metals,' *Trans Am. Inst. Min. Metall. Engrs.*, 1963, **227**, 1226—1228.

24b. S. Shiraisihi and R.G. Ward: 'The Density of Nickel in the Superheated and

Supercooled Liquid State,' *Canad. Metall. Quart.*, 1964, **3**, 117–122.

25. C. Ronchi, JRC Karlsruhe Establishment, private communication, 1991.

26. J.L. Margrave: 'Thermodynamic Properties of Liquid Metals', *High Temp – High Pressure*, `1970, **2**, 583–586.

27. G. Betz and M.G. Frohberg: 'The Enthalpy of Solid and Liquid Niobium,' *Scripta Metall*, 1981, **15**, 269–272.

28. V.Y. Chekovskoi, A.E. Sheindlin and B.Y. Berezin: 'Enthalpy Measurements by Drop Calorimetry Using Electromagnetic Levitation,' *High Temp – High Pressure*, 1970, **2**, 301–304.

29. J.P. Heirnaut and C. Ronchi: 'Calorimetic Measurements with Acoustic Levitation in High Temperature Heating Experiments with Pulse Laser Heating', *High Temp – High Pressures*, 1989, **21**, 119–130.

30. G.P. Hansen, S. Krishnan, R.H. Hauge and J.L. Margrave: 'Measurement of Temperature and Emissivity of Specularly Reflecting Glowing Bodies', *Metall. Trans.* 1988, **19A**, 1889.

31. S. Nakamura, T. Hibiya and F. Yamamoto: 'Effect of Convective Heat Transfer on Thermal Conductivity Measurement Under Microgravity Using a Transient Hot Wire Method, *Microgravity Sci. Technol*, 1992, **3**, 156–159.

32. G. Frohberg, K.H. Krantz and H. Weaver: 'The Diffusion and Transport Phenomena in Liquids under Microgravity,' *Proc. 6th Europ. Conf. on Mater. Sci. Under Microgravity*, Bordeaux, 1986, ESA SP 256, p.p. 585–591.

33. D.M. Herlach, R.F. Cochrane, I. Egry, H.J. Fecht and A.L. Greer: 'Containerless Processing of Metals and Alloys on Material Science in Microgravity,' *Intl. Mater. Reviews* (in press).

34. R.K. Wunderlich and H.J. Fecht: 'Specific heat Measurements by Non-contact Calorimetry,' *J. Non-Cryst. Solids*, 1993, **156-8**, 421–424.

35. T. Iida and R.L. Guthrie. *The Pysical Properties of Liquid metals*, Clarendon Press, Oxford, 1988.

36. I. Egry, B. Feuerbacher, G. Lohofer and P. Neuhaus: 'Viscosity Measurement in Undercooled Metallic Melts,' *Proc. 7th Europ, Symp. on Mater. Sci Under Microgravity*, ESA-SP 295, 1989, p.p. 257-260.

37. K.C. Mills and N. Machingawuta: 'SRM for High Temperature Viscosity Measurements. Results from phase 2 of interlaboratory Comparison Programme,' *NPL Report DMM(A)*, **29**, 1991.

# Plasma Enhanced Reactions

I.D. SOMMERVILLE AND A. MCLEAN

*Department of Metallurgy and Materials Science, University of Toronto, Toronto, Ontario, Canada.*

**Abstract**

The use of a plasma, generated by use of a drilled graphite electrode, has been found to enhance both the thermodynamics and kinetics of slag-metal transfer reactions. When the plasma employed is D.C., with reverse polarity (anodic electrode), still further enhancement can be achieved because of the superimposition of an electrochemical driving force onto the thermodynamic driving force.

Examples of this technology from the fields of both refining and smelting will be discussed; namely the separation of unwanted residual elements from the metal into the slag phase, and the recovery of metal values from slags and other oxide materials by reductive alloying into the Metal phase.

## Introduction

Most slag–metal reactions in steelmaking and refining are electrolytic in nature. This is not surprising since the components in a slag are predominantly, if not exclusively, in the ionic state. This fact has been established in many different investigations over the last 40 years or so, and our present understanding of the structures and properties of slags is firmly based in this concept. However, its repercussions with regard to slag–metal reactions have gone largely unnoticed. The main reason for this is that the two half reactions must proceed simultaneously to maintain charge neutrality, and usually they occur at the same location, the slag–metal interface. Hence, the overall reaction does not appear to be electrolytic, and the fact that electron transfer was involved tends to be forgotten. Under certain conditions, however, half reactions can proceed independently at different locations, and in this case the electrolytic nature of reactions becomes obvious.

The only reaction for which the electrolytic aspect has been explored in any depth is desulphurisation. [1-5] These studies have been discussed elsewhere, [6,7] and are briefly reviewed in a later section of this paper. However, the general point to be made at this juncture is that with the recent introduction of commercial D.C. arc furnaces, D.C. plasma furnaces, there is a new impetus to evaluate the effects of electrolysis in steel melting and refining.

With the advent of low cost solid rectifiers, high power D.C. arc systems have become practical. A few D.C. steel scrap melting furnaces have been brought on line in the last few years, and are performing well. [8] The main advantage of the D.C. scrap melting furnace graphite consumption. The wear rate of a D.C. negative electrode is about half that of the equivalent A.C. electrode and, by inference, about one quarter that of the equivalent D.C. positive electrode, predominantly due to

241

the importance of the 'burn rate' or tip loss in the high power melting furnace. In the lower powered ladle furnace, the electrode consumption is much less and is not a significant cost in the ladle refining process. Thus, although the D.C. positive mode of operation is not advisable in a high powered melting furnace, it may be useful and cost effective in ladle refining.

As ladle size diminishes it becomes increasingly more difficult to prevent damage to the refractory lining in a 3-electrode ladle furnace, even with argon injection. For this reason, the ladle furnace has not seen much use in foundry steelmaking. However, the single electrode D.C. plasma arc is a viable alternative, and has recently received attention by the foundries. A similar system may also be useful in tundish heating, especially for long duration casts such as in horizontal continuous casting. The flexibility of negative or positive polarity plasma heat should create interesting refining and alloying opportunities. However, an understanding of the electrolysis effects will be necessary to make use of these opportunities.

The use of a plasma, generated by the use of a drilled graphite electrode, has been found to enhance both the termodynamics and kinetics of slag–metal transfer reactions due to the higher temperature, fluidity and agitation of the slag phase. When the plasma employed is D.C., the use of reverse polarity (anodic electrode and cathodic melt), while having the disadvantage of higher electrode consumption, allows still further enhancement of these reactions to be achieved, due to the superimposition of an electrochemical driving force onto the thermodynamic driving forces for the appropriate reactions. This can be utilised to facilitate the separation of unwanted residual elements from the metal into the slag phase, and

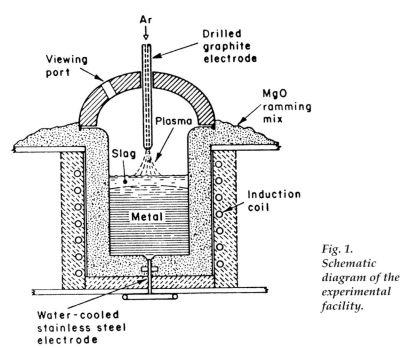

Fig. 1. Schematic diagram of the experimental facility.

also the recovery of metal values from slags, ores and other oxide materials by reductive alloying into the metal phase, as will be discussed in the following sections. A schematic diagram of the equipment is shown in Fig. 1. The experimental aspects have been dealt with elsewhere.[6,7]

## Enhancement of Refining Reactions

*Behaviour of Sulphur*
The overall slag–metal interfacial exchange reaction:

$$\underline{S} + (O^=) = (S^=) + \underline{O} \tag{1}$$

where the underlined components are in liquid steel and the bracketed components are in molten slag, can broken into its component half reactions:

$$\underline{S} + 2e = (S^=) \tag{2}$$
$$(O^=) = \underline{O} + 2e \tag{3}$$

Thus desulphurisation is cathodic in nature, and electroneutrality is maintained by anodic reactions such as reaction (3). Other possible anodic reactions in the steel/slag system may be:

$$\underline{Fe} = (Fe^{2+}) + 2e \tag{4}$$
$$\underline{Mn} = (Mn^{2+}) + 2e \tag{5}$$
$$\underline{Si} = (Si^{4+}) + 4e \tag{6}$$

The comparison between the desulphurisation achieved using cathodic and anodic melts (i.e. positive and negative top electrodes respectively) is shown in Fig. 2. Heat transfer at the cathode of a plasma is greater that at the anode. Consequently the melting rate of slag is faster with a negative electrode. This effect is apparent in Fig. 2, in that the initial desulphurisation rate is lower with the positive electrode than with the negative electrode, despite the greater overall rate.

As illustrated in Fig. 3, increasing plasma power results in improved desulphurisation rate and decreased final sulphur content of the melt. With the constant electrode to slag spacing employed, more power leads to higher voltage and current. The increased current entails higher current densities at the slag–metal interface which, in agreement with the findings of previous work,[2-5] results in improved desulphurisation. Also to advantage, the elevated slag temperature achieved with more plasma power gives rise to higher reaction rates and improved mixing.

It is conceivable that a high potential gradient through the slag may also be a factor in improved desulphurisation. The higher voltage drop would tend to polarise the slag and provide a driving force for sulphur anions to move through the slag away from the melt. It is also surmised that a sufficient potential gradient creates a

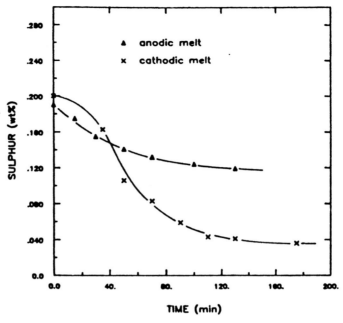

*Fig. 2. Variation of the sulphur content of anodic and cathodic melts with time.*

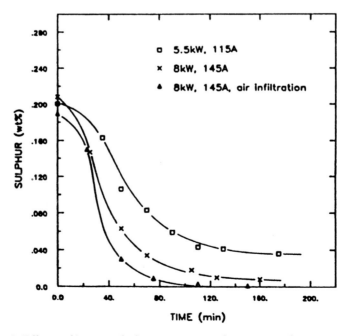

*Fig. 3. Effects of increased plasma power and current, and oxygen content in the gas phase on desulphurisation.*

slag which is more basic on the melt side than on the plasma side, thereby encouraging desulphurisation of the slag as well as the melt.

When a positive electrode (i.e. cathodic melt) is employed, reduction reactions occur at the slag–metal interface and oxidation reactions at the slag–plasma interface. Thus, oxygen and sulphur could be simultaneously removed at the slag–metal interface, with reaction (2) proceeding to the right and reaction (3) proceeding to the left. Furthermore, reactions (4) to (6) could also proceed to the left at the same time. However, the greater the number of reduction reactions taking place, the more slowly each proceeds since each is then consuming a fraction of the available Faradaic current. Thus, desulphurisation was much faster when the FeO and MnO contents of the slag and the oxygen activity in the steel were low. Interestingly, not only was sulphur removed more rapidly from the steel, but the slag sulphur content in the plasma impingement zone also decreased as time progressed, as shown in Fig. 4. The removal of sulphur from the slag to the gas phase during positive

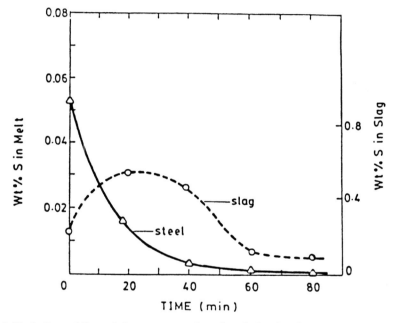

*Fig. 4. Variation of the sulphur content of steel and slag in plasma impingement zone with time (cathodic melt).*

electrode plasma heating can be attributed to two main factors. Firstly, the potential gradient and current through the slag are such that oxidising reactions are encouraged at the slag/plasma interface. Sulphur can therefore be removed by a reaction such as:

$$S^= + O_2 = SO_2 + 2e \qquad (7)$$

The voltage gradient may also cause the slag to polarise, with cations concentrating near the melt and anions concentrating near the gas phase, thereby increasing the activity of sulphur anions, and forcing reaction (7) to the right. Secondly, the high temperature and high gas velocity imparted by the plasma increase mass transport and reaction rates in both the slag and gas phases.

After desulphurisation of a cathodic melt, the typical average slag sulphur content was 0.6 wt.%. Local sulphur concentrations in the slag varied from an average of 0.3% in the region of plasma impingement to as high as 1.4% near the furnace wall. It follows that slag desulphurisation by the gas phase is greatest in the plasma impingement zone since both current density and temperature decrease rapidly outside this region. Sulphur lost to the gas phase amounted to 0–4% for anodic melts (electrode negative) and 50–80% for cathodic melts (electrode positive).

In agreement with reaction (7), it was found that increasing oxygen content of the gas phase led to a higher desulphurisation rate and a lower final sulphur level, as shown in Fig. 3. This result may seem contradictory to the established relationship between steel desulphurisation and oxygen potential, and it should be clearly understood that it applies only to the case where the anodic and cathodic reactions sites are well separated so that the slag is acting as a transfer medium or pump rather than as a reservoir for the sulphur. The corollary of this is that the sulphide capacity and hence the basicity of the slag become less critical.

Trials conducted in a 70 kW induction furnace at Mercury Marine in Wisconsin have confirmed the excellent refining capacity of the D.C. electrode positive plasma system. Using the commercial alloys CN7M, Monel 35-2, HY-80 and HP, the final sulphur levels obtained were 4, 4, 9 and 3 ppm respectively.[9]

After melting the charge in the induction furnace, the dissolved and total oxygen contents were quite high. The addition of aluminium and a basic slag cover reduced both values rapidly. However, during the period of negative electrode heating the total oxygen content slowly rode, presumably due to oxidation reactions at the slag–metal interface. On reversing the polarity, both the total and dissolved oxygen contents decreased, due to reduction at the slag–metal interface prevalent in this mode of D.C. plasma heating, as indicated by reaction (3).

## Enhancement of Smelting Reactions

*Slag Chemistry*

The influence of the direction of current flow on slag chemistry is evident in Table 1. An anodic melt increases concentrations of FeO, MnO and $Cr_2O_3$ in the slag by oxidising metallic elements in steel. Conversely, a cathodic melt causes reduction of these oxides, as well as of $SiO_2$. If the slag contains reducible oxides, then the fraction of the Faradaic current available for reduction of sulphur will be decreased. Consequently, the rate of desulphurisation of the melt is reduced, as is evident in Fig. 5. Reducible oxides were built up over time by plasma heating with a negative electrode, and then the current direction was reversed. The rate of desulphurisation during positive electrode heating was depressed more with a longer duration of prior negative electrode heating. It stands to reason that a higher activity of reducible oxides in the slag would utilise a larger fraction of Faradaic current, thus curtailing the desulphurisation rate.

| Slag | CaO | Al$_2$O$_3$ | MgO | Cr$_2$O$_3$ | FeO | MnO | SiO$_2$ | S |
|---|---|---|---|---|---|---|---|---|
| A | 46.6 | 46.6 | 1.9 | —— | 0.90 | —— | 3.4 | 0.22 |
| B | 44.11 | 43.80 | 9.11 | 0.50 | 0.25 | 0.20 | 1.87 | 0.56 |
| C | 41.13 | 40.31 | 5.10 | 5.35 | 1.47 | 2.78 | 3.36 | 0.97 |
| D | 42.45 | 41.97 | 8.12 | 2.25 | 1.14 | 0.99 | 2.52 | 0.73 |

Table 1. Typical slag compositions (wt%):

*A, artificial flux added (Alu-Cal; B, after 140 mins of desulphurisation with cathodic melt; C, after 135 mins desulphurisation with anodic melt; D, slag (C) after an additional 65 mins desulphurisation with cathodic melt.*

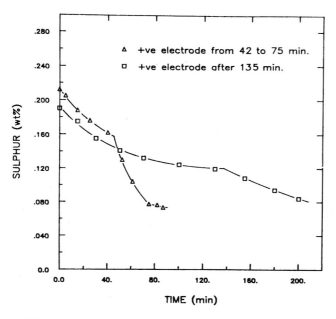

*Fig. 5. Effect of prior negative electrode plasma (anodic melt) duration on subsequent desulphurisation with positive electrode plasma (cathodic melt).*

247

*Reductive Alloying*

Obviously, the same polarity which favours Reaction (2) should favour other cathodic reactions such as the reverse of Reactions (4) to (6), which are oxide reduction or 'reductive alloying' reactions. As indicated above, this type of reduction of highly reducible oxides has been observed as a 'side effect' during electrochemically-enhanced desulphurisation experiments, but it is potentially of great importance in its own right, because it could prove to be extremely beneficial in several existing processes within the steelmaking and foundry industries, as well as having rather more speculative applications in the field of 'in-bath smelting' or 'smelting reduction'.

The superimposition of the electrochemical driving force on top of the normal thermodynamic driving force could enhance not only the reduction of iron oxide but also facilitate the addition of alloying elements such as nickel, manganese and possibly chromium, The addition of such elements as their oxides, followed by their reduction *in situ*, could by-pass the need for expensive ferroalloys (especially attractive where the desired carbon content of the melt is low) and thus significantly reduce the cost of allying. In addition, it could permit the use of indigenious, relatively low-grade ores such as chromites and reduce the amount of strategic materials such as low carbon ferrochrome which have to be imported.

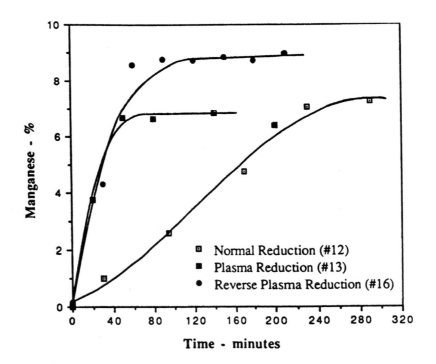

*Fig. 6. Effect of electrode polarity at a manganese to carbon ratio of one on the plasma reduction of manganese dioxide.*

Accordingly, experiments have been conducted with nickel oxide, manganese dioxide and chromium oxide in an apparatus very similar to that used for the desulphurisation studies, employing an induction furnace, a graphite crucible and a vertical, drilled graphite electrode. The oxides were pelletised with carbon and added to the top surface of a molten bath of carbon-saturated iron. The conditions employed were induction power designated only as 'normal reduction', plasma with normal polarity, i.e. anodic bath, designated as 'plasma reduction' and plasma with reverse polarity, i.e. cathodic bath, designated as 'reverse plasma reduction'.

For manganese oxide reduction, the best results were achieved at an manganese to carbon ration of one, which, since the oxide was $MnO_2$, corresponds to an oxygen to carbon ratio of two. It appears that since carbon is also available in the melt and from the crucible, provision of the full stoichiometric amount for CO formation is not required. In fact, excess carbon may physically clog up the melt surface, reducing the area available for oxide–melt interaction. The difference between normal and reverse polarity is shown on Fig. 6, the 'reverse plasma reduction' giving appreciably better recovery of manganese.

*Practical Implications*

This technology is particularly applicable in the foundry industry where induction melting is widely employed as the primary method of producing liquid metal. The ability to carry out reductive alloying under appropriate conditions with the metal stirred inductively and the slag superheated by the plasma power could provide and attractive alternative to the AOD process, by significantly increasing the range of ally steels which can be made by the induction melting route, without massive capital expenditure. It is also conceivable that it could provide a much less expensive route for the manufacture of nickel-base alloys and perhaps even cobalt-base alloys.

In the field of ladle metallury, arc heating is commonly used for raising and controlling temperature. The use of reverse polarity D.C. arcs would not only provide the necessary heating but also permit significant amounts of alloying subsequent to deoxidation and desulphurisation, using the relevant oxides rather than ferroalloys, under conditions of low oxygen potential and moderate stirring levels in the steel bath. Such conditions significantly reduce the loss of valuable alloying elements to the slag, as opposed to the oxidising environment which exists during tapping where much of the alloying is currently carried out.

Significant energy savings would accrue on the overall sequence of operations form raw materials to final product. In the current technology for production of alloy steels, ferro-alloys are produced at a separate site, usually by a separate company, and allowed to cool and solidify. These are then crushed to a size suitable for supply to the steel or foundry industries. In this new technology, the ferro-alloy would never exist as such. Instead, the reduction and alloying would occur simultaneously in the liquid state, so that the energy costs for crushing and sizing and later for reheating and melting of the ferro-alloy would all be eliminated. Energy would also be saved indirectly because of the higher and more predictable recovery of alloying elements mentioned earlier, and because of the elimination of the ferroalloys fines which inevitably from the crushing operation.

The D.C. plasma arc process has recently been implemented in the ladle at Maynard Steel Casting Company.[10] It has been demonstrated, as was predicted, [6,9] that the technique of argon injection through a single drilled graphite electrode is appropriate for use in the much smaller ladle sized (about 5 tons) typical of foundry operations, where the use of three electrodes would cause serious refractory damage. It can be used for reheating or superheating the steel, allowing longer holding times in the ladle and hence conferring much greater flexibility in the casting operations. This is particularly important for foundries since the heat losses from the smaller ladies are proportionately greater.

## Summary

By the use of D.C. electrode positive polarity, i.e. a cathodic melt or 'reversed polarity', together with a fluid, basic flux, the following benefits can be realised:

(a) Greatly accelerated sulphur transfer from metal to slag.

(b) Much lower sulphur contents in the melt that those dictated by chemical equilibrium, since up to 80% of the sulphur is removed from the slag–metal system by oxidation reactions at the anode.

(c) Very low oxygen activity and total oxygen contents, and hence low inclusion contents in the solidified steel, provided that the flux employed has high capacity for the deoxidation product (usually either silicated or alumina).

(d) Reductive alloying, which can provide alloying elements such as nickel, manganese and chromium by *in-situ* reduction of their oxides, and so reduce both alloy losses to the slag phase and the need for expensive ferro-alloy additions.

Thus, the significant electrolytic effects observed when using D.C. arc or plasma heating can confer major benefits by enhancement of slag–metal separation reactions which considerably outweigh the additional electrode costs involved.

## Acknowledgements

Financial support in the form of Operating Grants from the Natural Sciences and Engineering Research Council of Canada, for the desulphurisation studies, and from the Ontario Ministry of Energy for the investigation of reductive alloying, are gratefully acknowledged. The authors thank D. Frank Kemeny of Nupro Corporation and Mr. Mark Rishea of Ortech for their participation in the studies of desulphurisation and reductive alloying respectively.

# References

1.  S. Ramachandran, T.B. King and N.J. Grant: 'Rate and Mechanism of Sulphur Transfer Reaction,' *Trans. A.I.M.E.* 1956, **206**, 1549–1561.
2.  M. Ohtani and n.A. Gokcen: 'Effects of Applied Current on Desulphurisation of Iron,' *Physical Chemistry of Process Metallurgy*, Part II, G. St. Pierre, Interscience Publishers, new York, 1961, 1213–1227.
3.  R.G. Ward and K.A. Salmon: 'The Kinetics of Sulphur Transfer from Iron to Slag, Part II The Effect of Applied Current,' *Journal of the Iron and Steel Institute*, 1963, **201**, 222–227.
4.  P.M. Bills and R. Littlewood: 'Electrolytic Desulphurisation of Molten Iron,' *Journal of the Iron and Steel Institute*, 1965, **203**, 181–182.
5.  El-Gammal, T. B, Yostos and A Paksad: 'Indirect Electrolyte Desulphurisation' *Chemical Metallury of Iron and Steel*, Iron and Steel Institute, London, 1973, p.p. 144–147.
6.  F.L. Kemeny, I.D. Sommerville and A. McLean: 'Electrolysis Effects in D.C. Arc Processes,' *Proc, Electric Furnace Conference*, I.S.S.–A,I.M.E., 1989, **47**, 57–64.
7.  A. McLean, I.D. Sommerville and F.L. Kemeny: 'Enhanced Slag-Metal Reactions.' Proc, 4th Internat. Conference on Molten Slags and Fluxes, Iron and Steel Institute of Japan, Sendai, 1992 p.p. 268–273.
8.  D. Meredith and S.E. Stenkvist: 'Single Electrode D.C. Arc Furnace Operation at Florida Steel Corporation,' *Proc. Electric Furnace Conference*, I.S.S.–A.I.M.E., 1986, **44**, 43–47.
9.  F.L. Kemeny and L.J. Heaslip: 'Electrolysis Effects in D.C. Plasma Refining,' Paper presented at the Technical and Operating Conference, Steel Founders' Society of America, 1991.
10. J.A.T. Jones: D.C. Plasma Arc Furnace for Foundry Application, SME Technology Update for Casting, Chicago, September 1993.

# Kinetics of Hot Metal Dephosphorisation Using Lime Based Slags.

BRIAN MONAGHAN, ROGER POMFRET AND KENNETH COLEY

*Department of Metallurgy and Engineering Materials, University of Strathclyde, Glasgow, U.K.*

## Abstract

An investigation has been carried out to assess the effects of lime-based slags containing $SiO_2$, $CaF_2$ and $Fe_2O_3$ on hot metal dephosphorisation at 1330°C. It was found that the kinetics of dephosphorisation is first order with respect to phosphorous in the metal, and was controlled by mass transport in the slag. The oxygen potential that controls the phosphorous partition could be calculated from the FeO activity in the slag and optimum dephosphorisation was achieved with 50% $Fe_2O_3$ in the slag. Increasing the $CaF_2$ beyond 12.5% reduces the rate of dephosphorisation and lessens the amount of phosphorus removed.

## Introduction

The increase in demand for low phosphorous steels coupled with an increase in the phosphorous levels of the raw materials used in steelmaking has given the impetus for research into the removal of phosphorus from iron and steel.

The removal of phosphorous from iron and steel is an oxidation reaction needing a basic slag. Slags used for dephosphorisation in general contain $Na_2O$ or CaO. $Na_2O$-based slags are more powerful dephosphorising agents than CaO-based slags, but the CaO-based slags are preffered because of the process and environment problems involved in using $Na_2O$.[2]

The majority of work carried out on dephosphorisation has involved equilibrium studies of the phosphorus partition between slag and metal and there is only limited information on the kinetics of dephosphorisation.[3-8]

This investigation was set up in association with British Steel, Ravenscraig primarily to evaluate slags in the $CaO-SiO_2-Fe_2O_3$ system, some with the addition of CaF2, suitable for hot metal dephosphorisation in the convertor, and to obtain kenetic data on dephosphorisation using these slags.

## Experimental

Slag constituents, CaO, $SiO_2$ and $CaF_2$, were weighed and mixed. This mixture was then melted in an induction heated graphite crucible, then poured into a steel mould and allowed to cool to room temperature. The fused glassy slag was crushed and mixed with haematite to give a slag of the desired composition.

253

5kg of Sorel iron, with ferrophosphorus added to give the desired starting phosphorus level, was melted in a graphite crucible (inner diameter: 11 cm, height: 21 cm) in an induction furnace. It was raised to 1330°C (+/-10°) and a metal sample was taken. 500 g of prefused slag was added to the surface of the metal. Alternate metal and slag samples were taken every minute for 30 mins. The metal samples were obtained by suction using a silica tube and analysed for phosphorus using vanadium molybdate colorimetry. The slag samples were taken by means of a steel scoop and analysed by atomic absorption spectroscopy.

## Results and Discussion

*Kinetics of Dephosphorisation*
Figure 1 shows the results for experiments with varying initial phosphorus levels, and it can be seen that the rate of phosphorus removal increases with increasing initial phosphorus content of the metal. The result were tested for zero, first and second order kinetics with respect to phosphorus in the melt.

As can be seen in Fig. 1, increasing the starting phosphorus increases the rate. Therefore dephosphorisation is not zero order with respect to phosphorus in the metal.

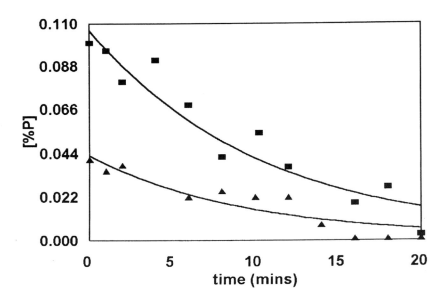

**Figure 1.** *Effect of initial [%P]*

First order dephosphorisation with respect to phosphorus in the metal can be represented by the following equation:

$$\frac{[\%P] - [\%P]_e}{[\%P]_0 - [\%P]_e} = \exp(kt) \tag{1}$$

where $[\%P]_0$ is the initial %P, $[\%P]_e$ is the equilibrium %P, $k$ is a first order rate constant, and $t$ is the time.

If dephosphorisation was first order then the likely rate limiting steps could be one or more of the following reactions:

1.  Chemical reaction at the slag metal interface.

2.  Mass transport control in the metal phase.

3.  Mass transport control in the slag phase.

This model gave a good fit to the experimental results as seen in Fig 1, where the solid lines are the best fit non-linear least squares regression lines.

Although not shown in this paper the results were also tested for second order behaviour but this model gave a very poor fit to the experimental data.

This is in good agreement with other reported work. Mori et al. [3-5] proposed a first-order two-film model based on both mass transport control in the metal and slag whilst Kitamura et al.[7] and Robertson et al.[8] proposed a first-order coupled reaction model based on mass transport control.

In order to investigate further, the rate controlling step experiments were carried out using different slag volumes which should establish whether or not mass transport in the slag is the rate controlling step.

For a first-order reaction controlled by mass transport in the slag Eq. (1) becomes:

$$\frac{[\%P] - [\%P]_e}{[\%P]_0 - [\%P]_e} = \exp\left( - \frac{[\%P]_0}{[\%P]_e} \frac{1}{h_s} k_s t \right) \tag{2}$$

where $h_s$ is the slag height and $k_s$ is the mass transfer coefficient for $PO_{2.5}$ in the slag.

Taking logs of Eq. (2) we obtain the following:

$$\ln \frac{[\%P] - [\%P]_e}{[\%P]_0 - [\%P]_e} = - \frac{[\%P]_0}{[\%P]_e} \frac{1}{h_s} k_s t \tag{3}$$

or

$$\frac{[\%P]_e}{[\%P]_0} h_s \left( \ln \frac{[\%P] - [\%P]_e}{[\%P]_0 - [\%P]_e} \right) = - k_s t \tag{3b}$$

The left hand side of Eq. (3b) is plotted against time in Fig. 2. for the experiments using different slag heights. The terms, $A$ and $Y$ in this figure are defined as follows:

$$A = -h_s \frac{[\%P]_e}{[\%P]_0} \qquad Y = \frac{[\%P] - [\%P]_e}{[\%P]_0 - [\%P]_e}$$

If mass transport in the slag was rate Controlling we would observe a single straight line relation to fit all of the experiments. Some difficulty was experienced in fitting the data to a single relation as different experiments showed different incubation times, related to the time required to form a slag. In order to overcome this problem the incubation time for each experiment was subtracted from the reaction time and this corrected time plotted in Fig. 2. The experiments show a reasonable fit to a single relation, suggesting that dephosphorisation is controlled by mass transport in the slag.

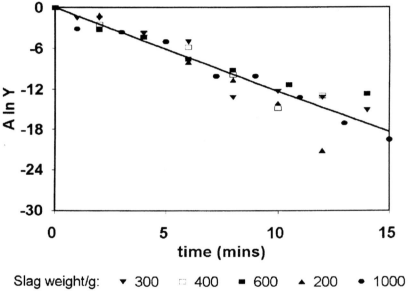

**Figure 2.** *First order rate plot for dephosphorisation, corrected for changing slag volume and incubation time.*

*Oxygen Potential*

A number of workers[9–11] have attempted, with limited success, to determine which reaction controls the oxygen potential and provides the driving force for dephosphorisation. In general it is accepted that the oxygen potential lies between that in equilibrium in the slag and that in equilibrium with the metal. Although this seems reasonable, it causes difficulty when attempting to calculate the expected phosphorus partition in a given situation.

Recently pan Wie *et al.*[2] attempted to tackle the problem of oxygen potential having found that neither the slag of the metal composition was in equilibrium with an oxygen potential consistent with the observed phosphorus partition. These workers proposed that, as the oxygen potential at the slag/metal interface was the result of a dynamic situation, it would not be in equilibrium withe the slag but is likely to be related to the slag composition. They demonstrated that the oxygen potential, calculated form experimentally observed partitions and calculated phosphate capacities, was a linear function of the $Fe^{3-}/Fe^{2-}$ ratio in the slag.

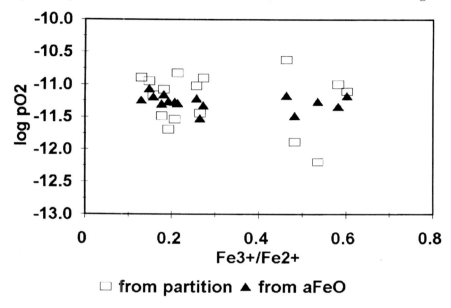

Figure 3. *Oxygen potential versus $Fe^3/Fe^2$ in the slag.*

Figure 3 shows oxygen potentials, calculated from phosphorus partitions observed in this work and phosphate capacities from the relation of Mori.[13] These results do not show any great dependency on $Fe^{3+}/Fe^{2+}$ ratio, although they do cover a fairly wide range of slag composition which was found by Pan Wei[12] to cause deviation from the straight line relation. The present authors have found that over limited ranges of slag composition the type of relation found by Pan Wei applies.

The reactions in a slag metal system that are likely to control the oxygen potential at the slag metal interface are.

(a)     $C + 0.5\,O_2 = CO$

(b)     $FeO = Fe + 0.5\,O_2$

Reaction (a) could be viewed as the metal controlling the oxygen potential, whereas (b) involves both phases and the oxygen potential would then be controlled by reaction at the slag/metal interface.

The oxygen potential was calculated for a number of experiments in this study, using each of the above reactions and the slag and metal compositions at the phosphorus reversion point. The FeO activities were calculated from the slag compositions using the I.R.S.I.D. computer model.

The oxygen potential calculated from the carbon oxygen reaction is much lower than expected from the observed phosphorus partitions, this finding is consistent with those of other workers. Oxygen potentials calculated using reaction (b) are also shown in Fig. 3 and are seen to be in good agreement with those calculated form the observed partition. This is to be expected, as oxidation of phosphorus would be expected to occur at the slag/metal interface, and therefore the reaction that controls the relevant oxygen potential should be one that occurs at that interface. What is more surprising is that the bulk slag FeO activity should give a reasonalbe interfacial oxygen potential when the slag is not in equilibrium with the metal. This observation would suggest that the FeO activity is uniform throughout the slag, and that there is an activity gradient of $Fe_2O_3$ in the slag. Therefore iron oxide reduction must be controlled by a mixture of mass transport in the metal and in the slag.

*Effect of $Fe_2O_3$*

Fig. 4 shows the effect of the $Fe_2O_3$ contentof the slag, and demonstrates that there is

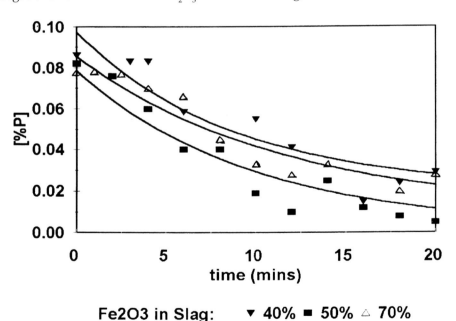

**Figure 4.** *The effect of slag iron oxide content on dephosphorisation*

not a simple linear relationship between the $Fe_2O_3$ content of the slag and the rate of dephosphorisation. Slower rates of dephosphorisation were obtained at 40% and 70% $Fe_2O_3$ compared to that of 50%. Also, lower final phosphorus levels were obtained using slags with 50 $Fe_2O_3$ suggesting some optimum composition in this region.

In Fig. 4 it can be seen that best dephosphorisation occurs at 50% $Fe_2O_3$. This is seen as and optimum level for dephosphorisation where the lower rates on either side of this composition can be explained as follows:

The 40% $Fe_2O_3$ slag is not sufficiently oxidising, thus limiting the slags ability to remove phosphorus down to a low a level, and slowing down the rate of phosphorus removal.

At the higher iron oxide level of 70% $Fe_2O_3$ there is a dilution effect, considerably reducing the amount of lime present in the slag. This would raise the final phosphorus content of the metal, as the slag would have a lower phosphorus capacity, and also reduce the rate of dephosphorisation.

50% $Fe_2O_3$ is seen as a compromise between these two conditions. Optimum levels of iron oxide for dephosphorisation have been reported by other workers. Balajiva et al.[14] obtained an optimum %FeO in slag for dephosphorisation of 14–16% at 1685°C. The different optimum levels of iron oxide in the slag are thought mainly to be the result of different working temperatures and the fact that Balajiva et al, were studying equilibrium partitions whereas the present work is concerned with a dynamic situation. In fact, the slag FeO content at the reversion point for the optimum slag compostion, 19%, is not very different from the optimum quoted by Balajiva et al.

*Effect of $CaF_2$*

Fig. 5 shown the effect of $CaF_2$ additions and shows that increasing the $CaF_2$ beyond

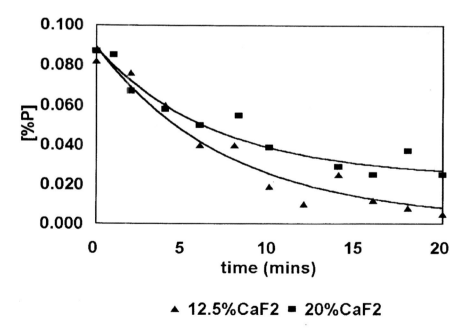

**Figure 5.** *Effect of $CaF_2$ on dephosphorisation.*

12.5% decreases both the rate of dephosphorisation and the amount of phosphorus removed.

Increasing the $CaF_2$ content of the slag beyond 12.5% decreased the amount and rate of dephosphorisation obtained and lessened the amount of phosphorus removed, as shown in Fig. 5. Ito and Sano,[9] who measured the effect of %$CaF_2$ on the phosphorus partition and the phosphate capacity, explained this by the fact that at high levels of $CaF_2$ the phosphate capacity iof the slag and the phosphorus partition are lowered, because $CaF_2$ has a dilution effect on the slag basicity. Simonov and Sano,[15] obtained similar results.

## Conclusions

It was found that the kinetics of dephosphorisation is first order with respect to phosphorus in the metal, and the rate was controlled by mass transport in the slag.

The oxygen potential that controls the extent of dephosphorisation is controlled by the equilibrium between FeO in the slag and iron in the metal.

Optimum dephosphorisation was achieved at 50% $Fe_2O_3$ in the slag.

Increasing the $CaF_2$ beyond 12.5% reduces the rate of dephosphorisation and lessens the amount of phosphorus removed.

## Acknowledgements

The authors would like to thank the Science and Engineering Research Council and British Steel PLC, Ravenscraig Works for financial support for this work.

## References

1.  K. Kunisada and H. Iwai:'Effects of CaO, MnO and $Al_2O_3$ on Phosphorus Distribution Between Liquid Iron and $Na_2O$-MgO-$Fe_1O$-$SiO_2$ Slags,' *Trans. ISIJ*, 187, 332.
2.  S.M. Cooper: 'Methods of Phosphorus Control Outside the Furnace,' *Proceedings of Phosphorus in Steel Conference*, University of Strathclyde, 1987, 65.
3.  K. Mori, S. Doi, T. Kaneko and Y. Kawai: 'Rate of Transfer of Phosphorus Between Metal and Slag,' *Trans. ISIJ*, 1978, 261.
4.  Y. Kawai, R. Nakao and K. Mori: 'Dephosphorisation of Liquid Iron by CaF2-base Fluxes,' *Trans. ISIJ*, 1984, 509.
5.  K. Mori, Y. Fukami and Y. Kawai: 'Rate of Dephosphorisation of Liquid Iron Carbon Alloys by Molten Slags,' *Trans. ISIJ*, 1988, 315.
6.  H. Suito and R. Inoue: 'Effects of $Na_2O$ and BaO Additions on Phosphorus Distribution between CaO-MgO-$Fe_1O$-$SiO_2$ Slags and Liquid Iron,' *Trans. ISIJ*, 1984, 47.

7. S. Kitamura, N. Sato and K. Okohira: 'Dephosphorisation and Desulphurisation of Hot Metal By CaO Based fluxes Containing Fe-oxide and Mn-oxide as Oxidant,' *Trans. ISIJ*, 1988, 364.

8. D.G.C. Robertson, B. Deo and S. Ohguchi: 'Multicomponent Mixed-Transport-Control Theory for Kinetics of Coupled Slag/Metal and Slag/Metal/Gas Reactions,' *Ironmaking and Steelmaking*, 1984, 41.

9. K. Ito and N. Sano: 'Phosphorus Distribution between Basic Slags and Carbon-saturated Iron at Hot Metal Temperatures,' *Trans. ISIJ*, 1985, 355.

10. H. Watanabe and K. Umezawa: 'Dephosphorisation with CaO based Flux and Estimation of Oxygen Potential at the Slag/Metal Interface,' *Tetsu to Hagane*, 1982, 517.

11. Y. Nakajima, M. Mukai, and N. Moriya: *Tetsu to Hagane*, 1982, 5954.

12. P. Wei, M. Ohya, M. Hirasawa, M. Sano, and K. Mori: 'Estimation of Slag Metal Interfacial Oxygen Potential in Phosphorus Reaction Between $Fe_1O$ Containing Slag and molten Iron with High Carbon Content,' *I.S.I.J. International*, 1993, 847.

13. T. Mori: *Trans. J.I.M.*, 1984, 761.

14. K. Balajiva, A.G. Quarrell and P. Vajragupta: 'A Laboratory Investigation of the Phosphorus Reaction in the Basic Steelmaking Process,' *J. Iron and Steel Inst.*, 1946, 115.

15. S. Simonov and N. Sano: 'Phophorus Equilibrium Distribution between Slags Containing MnO, BaO and $Na_2O$ and Carbon-saturated Iron for Hot Metal Pretreatment,' *Trans. ISIJ*, 1985, 1031.

# C. B. Alcock – List of Publications

## Section A: Thermodynamics of Metallic Solutions

1. 'Dilute Solutions in Molten Metals and Alloys', C.B. Alcock and F.D. Richardson., *Acta Met.* **6**, *385–395 (1958).*

2. 'Dilute Solutions in Alloys', *Acta Met.*, **8**, 882–887 (1960).

3. 'Some Problems in the Theory of Dilute Metallic Solutions', N.P.L. Symposium No. 9, *Phys. Chem. of Metallic Solutions and Intermetallic Compounds* (1959), p. 11.

4. 'Dilute Solutions of Sulphur in Liquid Tin and Lead', C.B. Alcock and L.L. Cheng, *Trans. AIME* **221**, 295–299 (1961).

5. 'Dilute Solutions of Sulphur in Liquid Iron, Cobalt and Nickel and Alloys Between These Metals', C.B. Alcock and L.L. Cheng. *J. Iron Steel Inst.* **195**, 169–173 (1960)

6. 'Thermodynamics and Solubility of Oxygen in Liquid Metals: I–Lead,' C.B. Alcock and T.N. Belford, *Trans. Far. Soc.* **60**, 822–835 (1964).

7. 'Thermodynamics and Solubility of Oxygen in Liquid Metals: Part II Tin', T.N. Belford and C.B. Alcock, ibid, **61**, 443–453 (1965).

8. 'Thermodynamics of Gold Alloys with Fe, Co, Ni', *J. Metals Science* **1**, C.B. Alcock and O.A.H. Kubik, 19–24 (1967).

9. 'Activities in Solid Binary Alloy Systems', A. Kubik and C.B. Alcock. *EMF Measurements in High Temperature Systems*, pp. 43–49, I.M.M. (1968).

10. 'Application of Thermodynamics to the Selection of Vapour Transport Reactions', C.B. Alcock and J.H.E. Jeffes. *Trans. I.M.M.* **C76**, 245–258 (1967).

11. 'The Production of Refractory Crystals by Vapour Transport Reactions', J.H.E. Jeffes and C.B. Alcock, *J. Materials Science* **3**, 635–642 (1968).

12. 'Thermodynamic Behaviour of Liquid Iron–Cobalt and Nickel–Platinum Alloys', C.B. Alcock and A. Kubik. *Trans, I.M.M.* **77**, C220–224 (1968).

13. 'The Formation of Alloys and Intermetallic Compounds by Vapour Phase Transport Reactions', C.B. Alcock and J.H.E. Jeffes, *Trans. I.M.M.* **77**, C195–200 (1968).

14. 'Solid Solubility of Carbon in Nickel', C.B. Alcock and G.P. Stavropoulos, *Trans. I.M.M.* **77**, C232–234 (1968).

15. 'A Thermodynamic Study of γ Phase Solid Solutions Formed Between Palladium, Platinum and Iron', C.B. Alcock and A. Kubik, *Acta Met.* **17**, 437–442 (1969).

16. 'A Mass Spectrometric Study of the Binary Liquid Alloys Ag–In and Cu–Sn', C.B. Alcock, R. Sridhar and R.C. Svedberg, *Acta Met.* **17**, 839–844 (1969).

17. 'A Thermodynamic Study of the Liquid Alloy Systems, Copper–Gallium and Copper–Germanium', C.B. Alcock and R. Sridhar and R.C. Svedberg, *J. Chem. Thermodynamics* **2**, 255–263 (1970).

18. 'Galvanic Cell Measurements of the Sodium–Oxygen System', C.B. Alcock and G. Stavropoulos, *Can. Met. Quart.* **10**, 257–265 (1971).

19. 'Quasichemical Equations for Oxygen and Sulphur in Liquid Binary Alloys', K.T. Jacob and C.B. Alcock, *Acta Met.* **20**, 221–232 (1972).

20. 'Thermodynamic Properties of Ag + Sn Alloys', P. Kubaschewski and C.B . Alcock, *J. Chem. Thermodynamics* **4** (1972).

21. 'Thermodynamics of Liquid Binary Alloys Between Copper, Silver and Tin, Germanium', P. Kubaschewski and C.B. Alcock, *Thermochimie* 211–218, Colloq. Intenationales CNRS, No. 201, Pans (1972).

22. 'The Prediction of Activities of Oxygen in Dilute Quaternary Solutions Using Binary Data', K.T. Jacob and C.B. Alcock, *Metallurgical Trans.* **3**,1913–1917 (1972).

23. 'Thermodynamics of α–Solid Solutions of Silver with Indium and Tin', C.B. Alcock, K.T. Jacob and T. Palamutcu, *Acta Met.* **21**, July, 1973, 1003–1009.

24. 'Activity of Indium in α–Solid Solutions of Cu + In, Au + In and Cu + Au + In Alloys', K.T. Jacob and C.B. Alcock, *Acta Met.* **21**, July 1973, 1011–1016.

25. 'Solute–Solute and Solvent–Solute Interaction in α–Solid Solutions of Cu + Sn, Au + Sn and Cu + Au + Sn Alloys', C.B. Alcock and K.T. Jacob, *Acta Met* **22**, May, 1974, 539–544.

26. 'Chemical Potential of Germanium in Its α–Solid Solutions With Copper, Silver and Gold', K.T. Jacob, C.B. Alcock and J.C. Chan, *Acta Met.* **22**, May, 1974, 545–551.

27. 'Solubility and Activity of Oxygen in Liquid Germanium–Copper Alloys', K. Fitzner K.T. Jacob and C.B. Alcock, *Met. Trans.* **8B**, 669 (1977).

28. 'Solubility and Activity of Oxygen in Liquid Gallium and Gallium–Copper Alloys', C.B. Alcock and K.T. Jacob, *J. Less–Common Metals* **53**, 211–222 (1977).

29. 'Determination of the Solubility of Oxygen in Liquid Indium by Combined Knudsen Mass Spectrometry and Coulometric Titration (in Japanese)', C.B. Alcock, E. Ichise and J. Butlert, *J. Japan Inst. Metals* **44**, 1239–1243 (1980).

30. 'Description of the Excess Thermodynamic Properties by the Chebyshev Polynomials', C.B. Alcock and V.P. Itkin in *Computer Modeling of Phase Diagrams*, 211–222 ed L.H. Bennett, Metallurgical Society, Inc. (1986).

31. 'The Au–Sr System', C.B. Alcock, V.P. Itkin, H. Okamoto and T.B. Massalski. 'The Ca–Sr System', C.B.Alcock and V.P. Itkin, *Bull. Alloy Phase Diagrams* **1** (5), 452–457 (1986).

32. 'The Au–Ba System', C.B.Alcock and V.P. Itkin. *Bull. Alloy Phase Diagrams* **7**(4), 336–337 (1986).

33. 'Thermodynamic Properties of the Fe–V System', V.P. Itkin, C.B. Alcock and J.F. Smith, *High Temperatures–High Pressures,* **18** (3), 271–276 (1986).

34. 'The Ba–Sr System', C.B. Alcock and V.P. Itkin, *Bull. Alloy Phase Diagrams* **8** (6), 534–536 (1987).

35. 'Thermodynamic Data for Uranium and Thorium Intermetallic Compounds: A Historical Perspective', C. B. Alcock, *J. Nuc. Mats.* **167**, 7–13 (1989).

36. 'The Al–Sr (Aluminum–Strontium) System', C.B. Alcock and V.P. Itkin, *Bull. Alloy Phase Diag.* **10**, 624–630 (1989).

37. 'The Si–Sr (Silicon–Strontium) System', V.P. Itkin and C.B. Alcock, *Bull. Alloy Phase Diag.* **10**, 630–634 (1989).

38. 'The Ca–Zn (Calcium–Zinc) System', V.P. Itlcin and C.B. Alcoek, *Bull. Alloy Phase Diag.* **11**, 328–333 (1990).

39. The Fe–Ni (Iron–Nickel) System, L.J. Swartzendruber, V.P. Itkin, and C.B. Alcock, *J. of Phase Equilibria,* **12**, 288–312 (1991).

40. 'The Ca–In (Calcium–Indium) System', H. Okamoto, V.P. Itkin, and C.B. Alcock, *J. of Phase Equilibria,* **12**, 379–383 (1991).

41. 'The Ba–Ga (Barium–Gallium) System', V.P. Itkin and C.B. Alcock, *J. of Phase Equilibria,* **12**, 575–577 (1991).

42. 'The Ca–Pb (Calcium–Lead) System', V.P. Itkin and C.B. Alcock, *J. of Phase Equilibria,* **13**(2), 162–169 (1992).

43. 'The Ga–Sr (Gallium–Strontium) System', V.P. Itkin and C.B. Alcock, *J. of Phase Equilibria,* **13**(2) 190–192 (1992).

44. 'The Ca–Ga (Calcium–Gallium) System', V.P. Itkin and C.B. Alcock, *J. of Phase Equilibria,* **13**(3), 273–277 (1992).


## Section B: Inorganic and Metallic Compounds  Thermodynamics

1.  'Ferrous Sulphide', C.B. Alcock and F.D. Richardson, *Nature* **168**, 661 (1951).

2. 'Sulphur Pressure Measurements Above FeS', C.L McCabe, C.B. Alcock and R.G. Hudson, *J. Metals,* 693–694, May (1956).

3. 'Metal Silicides and Silicon Carbide', P. Grieveson and C.B. Alcock, *Special Ceramics* **1**, 1983–2007 Heywood, London (1960).

4. 'Thermodynamics of the Gaseous Oxides of the Platinum Metals', C.B. Alcock and G.W. Hooper, *Proc. Roy. Soc.* (A), **254**, 551–561 (1960).

5. 'A Thermodynamic Study of the Compounds of Uranium with Silicon, Germanium, Tin, and Lead', C.B. Alcock and P. Grieveson, *J. Inst. Metals* **90,** 304–310 (1962).

6. 'Uraniurn Carbides and Borides', C.B. Alcock and P. Grieveson, *Thermodynamics of Nuclear Materials,* I.A.E.A. Symposium 563–579, Vienna (1962).

7. 'Analysis of Data Pertaining to the Vaporization of Uranium Carbides', C.B. Alcock, H.A. Eick, E.G. Rauh, R.J. Thorn, *Symposium on Carbides in Nuclear Energy,* AERE (1963).

8. 'Knudsen Effusion Studies of Compounds of Uranium and Thorium with Elements of Groups *IIIb and IVb'*, C.B. Alcock, J.B. Cornish and P. Grieveson, *Thermodynamics,* **1**, Int. Atomic Energy Agency, Vienna, 211–230 (1966).

9. 'The Free Energies of Formation of the Sulfates of Cobalt, Copper, Nickel and Iron', C.B. Alcock, K. Sudo and S. Zador, *Trans AIME* **233**, 655–661 (1965).

10. 'Free Energy Functions and Deviations in Thermochemical Data Storage for Inorganic Compounds', C.B. Alcock, *Nature* **209**, 198–199 (1966).

11. 'A Study of the Oxidation–Reduction Equilibria of Dilute Magnesiowustites', C.B. Alcock, and G.N.K. Iyengar. *Proc. Brit. Ceram. Soc.* **8**, 219–229 (1967).

12. 'A Thermodynamic Study of Dilute Solutions of Defects in the Rutile Structure $TiO_{2-x}$ and $Ti_{0.5}Nb_{0.5}O_{2-x}$', C.B.Alcock, S. Zador and B.C.H. Steele, ibid, 231–245, (1967).

13. 'Thermodynamic Study of the Manganese/Manganous Oxide System Using Solid Oxide Electrolytes', S. Zador and C.B. Alcock, *Electrochimica Acta,* **12**, 673–677 (1967).

14. 'Thermodynamique de la dissolution de l'oxygene et de l'azote dans le monocarbure d'uranium', C.B. Alcock, N.A. Javed and B.C.H. Steele, *Bull. Soc. Franc. Ceramic.* (77), 99–108 (1968).

15. 'Thermodynamic Study of $MoO_{2\pm x}$ With Small Deviations from Stoichiometry', S. Zador and C.B. Alcock, *J. Chem. Thermodynamics* **2**, 9–16 (1970).

16. 'The Corrosion of Ceramic Oxides by Astmospheres Containing Sulphur and Oxygen' C.B. Alcock and S. Zador, *Materials Science Research,* 9 (1971), Plenum Press, N.Y.

17. 'Thermodynamics of Uranium Oxycarbide Formation', B.C. H. Steel, N.A. Javed, and C.B. Alcock, *J. Nucl. Materials* **35**,1–13 (1970).

18. 'The Representation of Equilibria in Metal–Sulphur–Oxygen Systems by a Chemical Polential Diagram', *Can. Met. Quarterly* **10**, 287–289 (1971).

19. 'Solubilities of $Pb^{2+}$, $Bi^{3+}$, $Sb^{3+}$ and $Cu^+$ Sulphides in $NaCI + KCI + Na^2S$ Melts, O.H. Mohapatra, C.B. Alcock, and K.T. Jacob, *Met. Trans. AIME* **4**, 1955–62 (1973).

20. 'Electrochemical Measurement of the Oxygen Potential of the System Iron–Alumina–Hercynite in the Temperature Range 750–1600°C', J.C. Chan, C.B. Alcock and K.T. Jacob, *Can. Met. Quart.* **12**, No. 4, 439–443 (1973).

21. 'The Oxygen Potential of the Systems Fe + FeCr$_2$O$_4$+ Cr$_2$O$_3$ and Fe +FeV$_2$O$_4$ + V$_2$O$_3$ in the Temperature Range 750–1600°C', K.T. Jacob and C.B. Alcock, *Met. Trans.* **6B**, June 1975, 215–221.

22. 'Thermodynamics of CuAlO$_2$ and CuAl$_2$O$_4$ and Phase Equilibria in the System Cu$_2$O–CuO–Al$_2$O$_3$', K.T. Jacob and C.B. Alcock, *Amer. Ceram. Soc.* **8**, (5–6) 192–195 (1975).

23. 'Evidence of Residual Entropy in the Cubic Spinel Zr$_2$TiO$_4$', K. T. Jacob and C. B. Alcock, *High Temp–High Press.* **7**, 433 (1975).

24. 'Thermodynamics and Phase Equilibria in the System Cu$_2$O CuO–β–Ga$_2$O$_3$', K. T. Jacob and C.B. Alcock, *Revue Int. des Hautes. Temp.* **13**, 37 (1976).

25. 'Activities in the Spinel Solid Solution, Phase Equilibria and Thermodynamic Properties of the System Cu–Fe–O, K.T. Jacob, K.Fitzner and C.B. Alcock, *Met. Trans.* **8B**, 451 (1977).

26. 'Activities and their Relation to Cation Distribution in NiAl$_2$O$_4$ –MgAl$_2$O$_4$ Spinel Solid Solutions', K.T. Jacob and C.B. Alcock, *J. Solid State Chem.* **20**, 79 (1977).

27. 'Thermodynamic Properties of Fe$_3$0$_4$ – FeAl$_2$O$_4$ Spinel Solid Solution', A. Petric, K.T. Jacob and C.B. Alcock, *J. Amer. Ceram. Soc.* **64**, 632–639 (1981).

28. 'A Thermodynamic Study of Magneli and Point Defect Phases in the Ti–O Systems', S. Zador and C.B. Alcock, *High Temp. Sci.* **16**, 187–207 (1983).

29. 'The Control of Stoichiometry in Oxide Systems', C. B. Alcock, in *Non–Stoichiometric Compounds,* J. Nowotny and W. Weppner eds, Kluwer Academic Press, Dordrecht, 3–1 1 (1989).

30. 'The Vanadium–Oxygen System–A Review', C. B. Alcock and C. L. Ji, *High Temp.–High Press.* **22**, 139–147 (1990).

31. 'Strontium Oxide Activities in Oxide Ceramics by EMF Techniques', C. B. Alcock, *High Temperature Science* **28**, 183–187 (1990).

32. 'A Thermodynamic Study of the Cu–Sr–O System', C. B. Alcock and B. Li, *J. American Ceramic Soc.* **73**, 1170–1180, (1990).

33. 'A Thermogravimetric Analysis of Sr$_{14-x}$Ca$_x$Cu$_{24}$O$_{41}$', B. Li, C.B. Alcock, *Materials Letters,* **10**, 84–85 (1990).

34. 'Thermodynamics of the La–Sr–Cu–O High T$_c$ Superconductors, Part 1', R. Shaviv, E.F. Westrum, T.L. Yang, C.B. Alcock and B. Li, *J. Chem. Thermodynamics* **22**, 1025–1034 (1990).

35. 'Nonstoichiometry and Defect Characteristics of La$_{2x}$Sr$_x$CuO$_{4-x}$ B. Li, and C.B. Alcock, *Ceramic Transactions, Amer. Ceram. Soc.* **18**, 85–93 (1991).

36. 'Thermodynamics of La–Sr–Cu–O High–T$_c$ Superconductors', R. Shaviv, E.F. Westrum, C.B. Alcock and B. Li, *Superconductivity and its Applications, AIP Conference Proceedings 219,* Y.H. Kao , P. Coppens and H.S. Kwork eds, Amer. Inst. Physics, NY 200–205 (1991).

37. 'New Electrochemical Sensors for Oxygen Determination', C.B. Alcock, B. Li, J.W. Fergus and L. Wang, *Solid State Ionics* **53–56**, 3943 (1992).

38. 'Solid State Galvanic Oxygen Sensors (25–1600°C)', C.B. Alcock, B. Li, and L. Wang, *Proc. on Chemical Sensors II*, M. Butler, A. Ricco and N. Yamazoe eds , 93–97, 175 (1993).

## Section C: High Temperature Experimental Techniques

1. 'Radiochemical Measurements of High Temperature Equilibria lnvolving $H_2/H_2S$ Mixtures', C. B. Alcock, *JARI 3*, 135–142 (1958).

2. 'The Measurement of Vapour Pressures by the Transportation Method', C. B. Alcock and G. W. Hooper, *Phys. Chem. of Process Metallurgy,* Interscience, New York, 325–340 (1961).

3. 'The Vapour Pressures of Liquid Cu, Au and Ag', C. B. Alcock and P. Grieveson and G.W.Hooper, *Phys. Chem. of Process Metallurgy*, 341–352(1961).

4. 'Solid Oxide Electrolytes', C.B. Alcock and B.C.H. Steele, *Science of Ceramics,* Vol. II, 397–406, Academic Press (1965).

5. 'Factors Influencing the Performance of Solid Oxide Electrolytes', B.C.H. Steele and C.B. Alcock, *Trans. AIME* **233**.

6. 'Application of the Torsion Method in the Measurement of Vapour Pressures in the Temperature Range 1500–2500°C', M. Pollock and C.B. Alcock, *J. Sci. Inst.* **43**, 558–564 (1966).

7. 'Electrochemical Oxygen Sensors–Analysis and Modification', T.H. Etsell, S. Zador and C.B. Alcock, *Metal–Slag–Gas Reactions and Processes,* 834–850, Electrochem. Soc. Inc., Princeton, N.J. (1975).

8. 'Investigation of a New Technique for the Treatment of Steel Plant Waste Oxides in an Extended Arc Flash Reactor', C.A. Pickles, A. McLean, C.B. Alcock and R.S. Segsworth, *Advances in Extractive Metallurgy,* London (1977).

9. 'A New Route to Stainless Steel by the Reduction of Chromite Ore Fines in An Extended Arc Flash Reactor', *Trans. I.S.I. Japan* **18**, 369 (1978).

10. 'An Entropy Meter Based on the Thermoelectric Potential of a Nonisothermal Solid Electrolyte Cell', C.B. Alcock, K. Fitzner and K.T. Jacob. *J. Chem. Thermodynamics* **2**, 1011–1020, (1977).

11. 'Nonisothermal Probe for Continuous Measurement of Oxygen in Steel', T.H. Etsell, and C.B. Alcock. *Solid State Ionics* **314**, 621–626 (1981).

12. 'A Vapour Pressure Technique for the Determination of the Oxygen Transport Number of a Solid Electrolyte', C.B. Alcock, J. Butler and E. Ichise. *Solid State Ionics* **314**, 499–502 (1981).

13. 'Reduction of lron–Bearing Materials in an Extended Arc Flash Reactor', C. A. Pickles, A. McLean and C.B. Alcock. *Can. Met. Quanerly* **24**, 319–333 (1985).

14. 'Production of Ferronickel and Ferrovanadium from Fly Ash in an Extended Arc Flash Reactor', C.A. Pickles and C.B. Alcock, *Journal of Metals* **35**, 4045 (1983).

15. 'An Extended–Arc Reactor with Graphite Electrodes', C.B. Alcock, A. McLean and I.D. Sommerville. *MINTEK 50* Symposium proceedings, pp. 873–878 (1984).

16. 'Operation of a 1–MW Plasma Reactor/Zn–Pb Condenser Pilot Plant for Recovery of Metals from Arc-Furnace Dust', C. B. Alcock & I. Howitt, *Pyrometallurgy '87*, I.M.M. London (1987).

17. 'A Fluoride–based Composite Electrolyte', C.B. Alcock and B. Li, *Solid State Ionics* **39**, 245–249 (1990).

18. 'Electrochemical Experiments with Composite Electrolytes', C.B. Alcock, Proc. of the Discussion Meeting on Thermodynamics of Alloys, Barcelona, Spain, May 23–26, 1990, *Anales de Fisica* **86**, 87–91.

19. 'A Lanthanum–Fluoride-Based Composite Electrolyte for use as an Oxygen Sensor', C.B. Alcock and Li Wang, *High Temperatures–High Pressures* **22**, 449–451 (1990).

20. 'Electrochemical Studies with Fluoride Electrolytes', C.B. Alcock, *Pure & Appl. Chem.* **64**, 49–55 (1992).

21. 'Coupling Reactions on Oxide Solid Solution Catalysts', C.B. Alcock, J.J. Carberry, R. Doshi, and N. Gunasekaran, *Symp. on Natural Gas Upgrading II, Div. of Pet. Chem.*, ACS, San Francisco Mtg., April (1992).

22. 'The Preparation and Properties of Highly Conductive Nonstoichiometric Oxide Catalysts', C.B. Alcock and J.J. Carberry, *Solid State Ionics,* **50**, 197–202 (1992).

23. 'Perovskite Electrodes for Sensors', C.B. Alcock, R.C. Doshi and Y. Shen, *Solid State Ionics*, **51**, 281–289 (1992).

24. 'Electrochemical Sensors for the Non–Metallic Elements', C.B. Alcock, *Proc. of Sundaram Symp.* Kalpakkam, India, Nov., 1989 (1991).

25. 'The Electrolytic Properties of $LaYO_3$ and $LaAlO_3$ Doped with Alkaline–Earth Oxides', C.B. Alcock, J.W. Fergus and L. Wang, *Solid State Ionics*, **51**, 291–295 (1992).

26. 'Impedance Analysis of Solid Electrolytes by a Voltage–pulsing Technique', P.J. Komorowski and C.B. Alcock, *Solid State Ionics*, **58**, 293–301 (1992).

## Section D: Kinetics of High Temperature Processes

1. 'The Kinetics of Vaporization of Refractory Materials', C.B. Alcock and E. Wolff, *Trans. Brit. Ceramic. Soc.* **61**, 667–687 (1962).

2. 'A Study of Cation Diffusion in Stoichiometric $UO_2$ Using $\alpha$–Ray Spectrometry', with R.J. Hawkins, A.W.D. and P. McNamara *Thermodynamics Vol–II, International Atomic Energy Agency*, Vienna 1966, 57–72.

3. 'Kinetics of Sulfate Formation on Cuprous Oxide', M.G. Hocking and C.B. Alcock, *Trans. AIME* **236**, 635–541 (1966).

4. 'Kinetics and Mechanism of Cobalt Sulphate Forrnation', C.B. Alcock and M.G. Hoclcing, *Trans. IMM* C27–C36 (1966).

5. 'Vaporization Kinetics of Ceramic Oxides at Temepratures Around 2000°C', C.B. Alcock and M. Peleg, *Trans. Brit. Ceram. Soc.* **66**, 217–232 (1967).

6. 'A Comparison of the Calculated and Observed Vaporizalion Behaviour of Uranium Carbide', C.B. Alcock, H. Eick, E. Rauh and R.J. Thorn. *AIME Symposium*, 257–266 (1964),

7. 'La Vaporisation Sous Vide Des Oxydes Ceramiques', *Bull. Soc. Ceram. Francaise.* **72**, 2536 (1966).

8. 'A Study of Cation Diffsion in $UO_{2-x}$ and $ThO_2$ Using $\alpha$–Ray Spectrometry', R.J. Hawkins and C.B. Alcock, *J. Nuclear Matenals* **26**, 112–122 (1968).

9. 'The Corrosion of Nickel in $SO_2$ $O_2$ Mixtures in the Temperature Range 500–750°C', C.B. Alcock, M.G. Hocking and S. Zador, *J. Corrosion Science* **2**, 111–122 (1969).

10. 'Physico–Chemical Factors in the Dissolution of Thoria in Solid Nickel', C.B. Alcock and P.B. Brown, *J. Met Sci* **3**, 116–120 (1969).

11. 'A Study of Semi–Conduction in Dilute Magnesio Wustites', G.N.K. Iyengar and C.B. Alcock, *Phil. Mag* **21**, 293–304 (1970).

12. 'The Corrosion of Tantalum by $Na_2O$–Saturated Liquid Sodium', C.B. Alcock, M.G. Barker and G.P. Stavropolus, *Corrosion Science* **10**, 105–110 (1970).

13. 'Ionic Transport Numbers of Group IIA Oxides Under Low Oxygen Potentials', C.B. Alcock and G.P. Stavropoulos, *J. Amer. Ceramic. Soc.* **54**, 436–443 (1971).

14. 'Cation Diffusion in $UO_2$–based Solid Solutions as a Function of Stoichiometry, H.M. Lee and C.B. Alcock, *Proc. Brit. Ceramic. Soc.* **19**, 75–93 (1971).

15. 'Growth Kinetics of Dispersed $ThO_2$ in Nickel and Nickel–Chromium Alloys, P.K. Footner and C.B. Alcock, *Metallurgical Trans.* **3**, 2633–2637 (1972).

16. 'Electrolytic Removal of Oxygen from Gases by Means of Solid Electrolyte', C.B. Alcock and S. Zador, *J. Applied Electrochem* **2**, 289–299 (1972).

17. The Oxygen Permeability of Stabilized Zirconia Electrolytes at High Temperatures, C.B. Alcock and J.C. Chan, *Can. Met. Quart.* **11**, 559–567 (1972).

18. 'The Corrosion of Ceramic Oxides by Atmospheres Containing Sulphur and Oxygen', with S. Zador, *Matenals Science Res., 5 –Ceramics in Severe Environments*, W.W. Kriegal and H. Palmer, eds, III, Dec. 7–9, (1970), 1–10, Plenum Press.

19. 'The Mechanism of Vaporization and the Morphological Changes of Single Cryslals of Alumina and Magnesia at High Temperatures, M. Peleg and C.B. Alcock, *High Temperature Science* **6**, 52–63, (1974).

20. 'Particle Coarsening in Fused Salt Media. R.D. McKellar and C.B. Alcock, *Sintering and Catalysis*, ed. G.C. Kuczynski, 409–418, Plenum Press, N.Y. (1975).

21. 'Catalytic Oxidation of Carbon Monoxide over Superconducting $La_{2-x}Sr_xCuO_4$ Systems between 373–523 K', S. Rajadurai, J.J. Carberry, B. Li and C.B. Alcock, *J. of Catalysis*, **131**, 582–589 (1991).

22. 'Solid State Sensors and Process Control', C.B. Alcock, *Solid State Ionics*, **53–56**, 3–17 (1992).

23. 'Carbon Monoxide and Methane Oxidation Properties of Oxide Solid Solution Catalysts', R. Doshi. C.B. Alcock, N. Gunasekaran and J.J. Carberry, *J. of Catalysis*, **140**, 557–563 (1993).

24. 'Effect of Surface Area on CO Oxidation by the Perovskite Catalysts $La_{1-x}Sr_xMO_{3-\delta}$ (M=Co,Cr)', R. Doshi, C.B. Alcock and J.J. Carberry, *Catalysis Letters*, **18**, 337–343 (1993).

### Section E.  Survey Papers

1. 'The Formation of Volatile Oxides by Furnace Construction Materials', C.B. Alcock, *Trans. Brit. Ceram. Soc.* **60**,147–164 (1961).

2. 'Gaseous Oxides of the Platinum Metals', C.B. Alcock, *Platinum Metals Review* 134–149, October (1961).

3. 'The Experimental Development of Chemical Metallurgy', C.B. Alcock, *Chipman Conference* (1962), 29–37.

4. 'The Choice of Materials for High Temperature Operation', C.B. Alcock, and C.L. McCabe, ibid, (1962), 49–57.

5. 'Thermodynamics of Gas–Metal Systems', *Corrosion Newnes* (London) 7,120–7,138 (1962).

6. 'Equilibria and Transport at Ceramic Interfaces', C.B. Alcock, *Materials Science Research* **3**, 1–10 (1966), Plenum Press, New York.

7. 'Dilute Solutions in Liquid Metals and Alloys with Particular Reference to Oxygen and Sulfur', C.B. Alcock, Metallurgical Thermodynamics Symposium, Pittsburgh (1964).

8. 'Problems in the Kinetics of Reactions Involving Ionic Solids', C.B. Alcock, *Materials Science Research (2)*, ed. G.C. Kuczynski (1965).

9. 'Physico–Chemical Measurements at High Temperatures', C.B. Alcock and F.D. Richardson, *Chemical Equilibria*, Butterworths, 136–136 (I959).

10. 'The Applicability of Some Simple Models to Metallurgical Solutions', C.B. Alcock and R.A. Oriani, *Trans AIME* **224**, 1104–1114 (1962).

11. 'The Uranium–Carbon and Plutonium–Carbon Systems', C.B. Alcock as Member of IAEA Panel. *Techn. Rep. Series* **14**, 44 (1962).

12. 'Problems with Sollutions: A Chemical View of High Temperature Materials', Inaugural Lecture, Imperial College (1965).

13. 'Transport of Ions and Electrons in Ceramic Oxides' (Proc. of a Symposium, 'Electromotive Force Measurements in High–Temperature Systems', The Nuffield Research Group, Imperial College, April 13–14, 1967–Institution of Mining and Metallurgy, London, 109–124).

14. 'High Temperature Chemistry of Simple Metallic Oxides', *Chemistry in Britain 5*, 216–223 (1969).

15. 'Solution Thermodynamics in Metallic and Ceramic Solid Systems, *Ann. Rev. Mat. Sci.* **1**, 219–252 1971.

16. 'The Control of Chemical Composition in High Temperature Systems', Colloques Internationaux du Centre National de la Recherche Scientifique, No. 205, Etude des Transfonnations Cristallines a Haute Temperature au–Dessus de 2000 K–Odeillo, 27–30, September, 1971.

17. 'Aspects of Progress in the Science of Pyrometallurgy', *Met. Trans. B., Process Metallurgy*, **6B**, (1), March, 1975, (1974 Extractive Metallurgy Lecture, AIME).

18. 'Modern Aspects of Thermodynamics Applied to Metal Extraction', *Trans. Indian Institute of Metals* **27**, (4), August, 1974, 183–192.

19. 'The Oxygen Probe as a Ceramic System', *Sintenng and Catalysis*, G.C. Kuczynski ed. 419–434, Plenum Press, N.Y. (1975).

20. 'Plasma Processing of Oxide Systems in the Temperature Range l000–3000K', C.B. Alcock. *Pure and Appl. Chem.* **52**, 1817–27 (1980).

21. 'Chemical Aspects of the Choice of Secondary Melting Point Standards for the Temperature Range 2000–3000°C', C.B. Alcock, *High –Temperature–High Press.*, **II**, 241–49 (1979).

22. 'Basic Science and Process Optimization in Extractive Metallurgy', C.B. Alcock. *Trans. Indian Inst. Metals*, **32**, 195–199 (1979).

23. 'Solution Models and Patterns in Alloy Thermodynamics'. C.B. Alcock. *Pure and Appl. Chem.*, **55**, 529–538 (1983).

24. 'Materials Processing in Plasma Furnaces Equipped with Graphite Electrodes', I.D. Sommerville, A. McLean, C.B. Alcock and C.A. Pickles, *Plasma Technology in Metallurgical Processing*, J. Feinman. ed. 890–1001, Iron and Steel Society (1987).

25. 'Evaluating Thermodynamic Data for use in the Laboratory and Industry', C. B. Alcock, Toronto Meeting of American Chem. Soc. (1988).

26. 'High Temperature Sensors for NonMetallic Elements', C. B. Alcock, *CUICAC Workshop Reports*, 1–15 (1989).

27. 'Chemical Sensors for Pyrometallurgy', *Advanced Materials*, 15–20, V.I. Lakshmanan ed., Society for Mining, Metallurgy & Exploration, Colorado (1990).

28. 'Thermodynamic Properties of the Group IIA Elements', C.B. Alcock, M.W. Chase and V. Itkin, *Amer. Chem. Soc. and Amer. Inst. of Physics for N.I.S.T.* **22**, 1–85 (1993).

29. 'Electrochemical Oxygen Sensors', C.B. Alcock, *Rev. Int. Hautes Temper. Refract.* **28**, 1–8 (1992–93).

30. 'Thermodynamic and Transport Properties of Electrocerarnic Oxides Systems', C.B. Alcock, *J. Alloys & Compounds*, **197**, 217–227 (1993).

## Section F: Miscellaneous

1. 'Scattering of Neutrons by Deuterons', C.B. Alcock, J.L. Martin, E.H.S. Burhop and R.L.F. Boyd, *Proc. Phys. Soc.* A **63**, 884–897 (1950).

2. 'Phosphorus and Sulphur in Austenitic Stainless Welds', S.M. Makin, C.B. Alcock, D.R Arkell and P.C.L. Pfeil, *British Welding Journal*, 595–599, October (1960).

3. 'Some Considerations of the Theory of Growth of Graphite Nodules in Solidifying Cast–Iron Alloys', J.D. Schobel and C.B. Alcock, *Zeit. fur Metalkunde*, **57**, 74–76 (1966).

4. 'La Production des Cristaux Refractaries Par Les Reactions de Transports Chuniques', J.H.E. Jeffes and C.B. Alcock, *Bull. Ceram. Soc, Francaise* **80**, 29–44 (1968).

5. 'Grundlagenforschung und Industrielle Entwicklung in der Eisen-und Stahlindustrie', *Radex-Rundschau* **2**,117 (1977).

6. 'Zirconium: Physico–Chemical Properties of its Compounds and Alloys', Int. Atom. Energy Agency (1976).

7. 'Thermodynamics of the Titanium–Oxygen System in the Temperature Range 1300–1500°C', S. Zador and C.B. Alcock, *High Temp. Sci.* **16**, 187 (1983).

8. 'Continuous Oxygen Determination in a Concast Tundish', T.H. Etsell and C.B. Alcock, *Can. Met. Quarterly* **22**, 421(1983).

9. 'Some Observations on Activation of FeTi for Hydrogen Absorption', D. Khatamian, N.S. Kazama, G.C. Weatherly, F.D. Manchester and C.B. Alcock, *J. Less Common Metals* **89**, 781 (1983).

10. 'Measurement of the Magnetization of Activated FeTi, D. Khatanmian, F.D. Manchester, G.C. Weatherly and C.B. Alcock in *Electronic Structures and Propertes of Hydrogen in Metals*, P. Jena and C.B. Sattersthwaite eds. Plenum Press 737 (1983).

11. 'Smelting and Refining of Ferro–Alloys in a Plasma Reactor', I.D. Sommerville, A. McLean and C.B. Alcock, AIME Electric Furnace Conference, Detroit (1983).

12. 'Vapour Pressures of the Metallic Elements', C.B, Alcock, V. Itkin and M. Homgan, *Can. Met. Quaterly*, **23**, 309–313 (September, 1984).

13. 'Some Chemical Equilibria for Accident Analysis in Pressured Water Reactor Systems', P. E. Potter, M. H. Rand and C. B. Alcock, *Journal of Nuclear Materials* **130**, 139–153 (1985).

14. 'New Technology in the Smelting of Sulphides', C. B. Alcock and J. M. Toguri, *Report to Ministry of Energy*, Ontario, 43 (1984).

15. 'Relative Ionization Efficiencies from Alloy Thermodynamic Data', C. B. Alcock, *High Temperatures–High Pressures* **20**, 165–168 (1988).

16. 'Solid Oxide Solutions as Catalysts–A Comparison with Supported Pt', J. J. Carberry, S. Rajadurai, C. B. Alcock and B. Li, *Catalysis Letters* **4**, 43–48 (1990).

17. Handbook of Chemistry and Physics, 71st edn, 1990–1991, contribution by C. B. Alcock (1990).

18. 'X–ray Diffraction of Lanthanum Aluminate Doped with Alkaline–Earth Cations', J. W. Fergus and C. B. Alcock, *Materials Letters* **12**, 219–221 (1991).

**Section G: Books**

1. *Principles of Pyrometallurgy*, Academic Press, London, 1976.

2. *Metallurgical Thermochemistry* (with o. Kubachewski), Pergamon Press, London, new edition in preparation.

3. *Physical Chemistry of Nuclear Power*, in preparation.

4. *Electromotive Force Measurements in High Temperature Systems* (as editor), Institute of Mining and Metallurgy, London 1967.